SCIENCE ON TRIAL

SCIENCE ON TRIAL

The Whistle-blower, the Accused, and the Nobel Laureate

■

JUDY SARASOHN

St. Martin's Press New York

For my mother and in memory of my father

Design by Jaye Zimet

Library of Congress Cataloging-in-Publication Data

*Sarasohn, Judy.
Science on trial : the whistle-blower, the accused, and the Nobel laureate / Judy Sarasohn.
p. cm.
ISBN 0-312-09247-4
1. Cellular immunity—Research—Moral and ethical aspects.
2. Fraud in science. 3. Baltimore, David. 4. O'Toole, Margot.
5. Imanishi-Kari, Thereza. I. Title.
QR185.5.S37 1993 93-901
364.1'63—dc20 CIP*

First Edition: October 1993

10 9 8 7 6 5 4 3 2 1

It's a kind of scientific integrity, a principle of scientific thought that corresponds to a kind of utter honesty—a kind of leaning over backwards. For example, if you're doing an experiment, you should report everything that you think might make it invalid—not only what you think is right about it; other causes that could possibly explain your results; and things you thought of that you've eliminated by some other experiment, and how they worked—to make sure the other fellow can tell they have been eliminated.

Details that could throw doubt on your interpretation must be given, if you know them. You must do the best you can—if you know anything at all wrong, or possibly wrong—to explain it. . . . In summary, the idea is to try to give all of the information to help others to judge the value of your contribution; not just the information that leads to judgment in one particular direction or another. . . . We've learned from experience that the truth will come out. Other experimenters will repeat your experiment and find out whether you were wrong or right. Nature's phenomena will agree or they'll disagree with your theory. And although you may gain some temporary fame and excitement, you will not gain a good reputation as a scientist if you haven't tried to be very careful in this kind of work.

—Richard Feynman, in *"Surely You're Joking, Mr. Feynman!" Adventures of a Curious Character*

Contents

Acknowledgments

■

When I began work on this book, scientists had already gone through five years of fighting over allegations by a former postdoctoral student that experiments had not been done as published in *Cell,* a scientific journal. They were wary of someone new asking questions. Are you a scientist? many demanded.

I explained to one longtime MIT supporter of David Baltimore that I was not a scientist, but that the story was such a compelling one about human frailties and strengths that I wanted to write about it for a much broader audience. You'll never understand the science, Baltimore's friend said. But when I asked if he would explain it, the scientist begged off. The science is complicated, he said.

Whom could I talk to about the science, someone neutral? I persisted.

There is no one neutral, this scientist said. Everyone is polarized.

His response spoke volumes about the dispute over the *Cell* paper. Many scientists had not bothered to, or did not want to, look at the actual paper and allegations in dispute, and their feelings about the controversy were so raw that they did not believe other scientists could be objective.

In the end, the story of Margot O'Toole, the whistle-blower; Thereza Imanishi-Kari, the accused; and David Baltimore, the Nobel laureate, was about much more than the science.

To get at that story, I interviewed nearly a hundred scientists, government officials, lawyers, and others concerned about scientific misconduct. I also pored over more than two thousand pages of transcripts of congressional hearings, laboratory records, official interviews with scientists involved in the controversy, letters, records of meetings, and the scientists' own written statements about the dispute.

Although I greatly appreciated the help of all the scientists who took my calls, I particularly want to thank Margot O'Toole for her time. O'Toole spoke with me even though she, unlike most other whistle-blowers I have interviewed, was reluctant to discuss the case yet again. Also especially patient and helpful were National Institutes of Health scientists Ned Feder and Walter Stewart; former NIH investigator Suzanne Hadley; and Representative John Dingell and his aide Peter Stockton.

I am also grateful to Thereza Imanishi-Kari and David Baltimore, who understandably could have refused to talk to me, but instead were generous with their time, as were Brigette Huber and Robert Woodland, members of the ad hoc Tufts review panel, and Imanishi-Kari supporters Joan Press and David Parker.

In addition, I must thank my agent, Jane Dystel, who was always sure there was a book that I should write about the controversy, and St. Martin's editor Bill Thomas, whose dedication to the project and enthusiasm never wavered. I thank as well Keith Kahla, who adopted my project at St. Martin's.

In my nonbook life, I work for *Legal Times,* a weekly Washington, D.C., newspaper that covers law and lobbying. It was for *Legal Times* that I first wrote about the Baltimore case in 1988, when a politically well-connected Washington law firm registered to lobby on Baltimore's behalf. Thanks go to my *Legal Times* editors Eric Effron and Ann Pelham, and to *American Lawyer*'s editor-in-chief, Steve Brill, who encouraged me and allowed me the time to work on the book.

And most of all, I want to express my deep appreciation to Kim Isaac Eisler, my husband and colleague, for his professional assistance and personal support, and to Sara Sophie Eisler, who shared her first two years with this book.

ONE

The Whistle-blower

■

In the spring of 1986, Margot O'Toole, a young, chestnut-haired woman whose speech had not lost the lilt of her native Dublin, was checking out the mice at MIT.

This was not the assignment O'Toole had imagined when she left the prestigious Institute for Cancer Research in Philadelphia. Here at the Center for Cancer Research at Boston's Massachusetts Institute of Technology, Margot had been anything but a star. The promise that had shone when her doctorate was awarded seven years earlier had faded. She lamented to friends that she was stumped and frustrated.

After initially getting some exciting results from an immunology experiment, the thirty-two-year-old researcher was stymied by a project based on the work of Thereza Imanishi-Kari, her new boss. Imanishi-Kari's work was soon to be heralded as an important breakthrough in the history of immunology. Imanishi-Kari's revolutionary discovery was called idiotypic mimicry. Her experiments indicated that when a foreign gene is introduced into a specially bred laboratory mouse, the mouse's own genes begin to produce antibodies of the type made by the foreign one. If these findings were correct, doctors might be able to introduce genes that would influence a person's own immune system to attack a disease. In the future, a body might be reprogrammed to do its own damage against cancerous cells. The study of immunology, as ambiguous a science as it was, was indeed at the cutting edge of medical research in the 1980s and 1990s.

For the moment, here in Cambridge in 1986, the hopes for the experiment were not immodest. Perhaps it would lead to a cure for such immune disorders as AIDS and lupus. The basic research had already begun to show results. All O'Toole had to do was repeat part of the experiment in order to extend her boss's work; then she could move forward to even bolder discoveries. Her work was supposed to show how idiotypic mimicry worked. But O'Toole could not get past the first step.

Margot was the daughter of a true Renaissance man. James D. B. O'Toole, an engineer by training, once worked for the Irish national power company, and had been a champion swimmer, rugby player, and discus thrower. But he was also a formidable journalist and writer; he even analyzed foreign affairs on Irish radio. While working for the power company, O'Toole had written a play that was later produced by the Olympia Theatre Company in Dublin. The play, which evinced a "no-man-is-an-island" theme, examined a man's relationship to his job. It caused quite a stir, and his employers were not happy about it.

There were three daughters and a son in the O'Toole family. All were imbued with a love of education, poetry, and athletics. Margot excelled at swimming, horseback riding, dance, and French. At five years of age, she recited from memory sophisticated poetry with an understanding and diction far beyond her years. In Ireland, said her mother, Elizabeth, "poetry was an everyday part of your upbringing."

The O'Toole household was often filled with relatives, friends, and colleagues, and the O'Toole children were privy to adult conversations about the news of the day. They also heard stories of their ancestors who had fought in the great Irish rebellion, and tried to save others from starving during the infamous famine. "There was," Mrs. O'Toole said, "a history of doing the right thing." Margot was a bright, engaged child, raised in an atmosphere of achievement and intellectual and ethical refinement.

James O'Toole came to America in 1964—his family followed three years later—after he was offered a one-year job teaching creative writing at Ohio State University. He later moved to Boston University, where he set up a program in science writing at the journalism school.

Margot's life path was set during her high school years, when she became "completely fascinated" by a biology class. "It was learning about life, how life worked," she recalled. This early fascination with the mechanics of life fueled her drive to become a scientist. She attended Brandeis University and then received her doctorate in immunology from Tufts in 1979, studying under Dr. Henry Wortis. While at Wortis's lab, she met Peter Brodeur, a handsome, six-foot-two-inch scientist who was Wortis's third doctoral student (O'Toole was his first). The two younger scientists fell in love and were married. Wortis, a friend as well as a mentor, was a guest at their wedding.

O'Toole then moved to Philadelphia to work at the Institute for Cancer Research, first as a postdoctoral fellow of the National Institutes of Health and then as a fellow of the National Arthritis Foundation. She was happy in Philadelphia, but agreed to leave when her husband took a position at the Tufts University School of Medicine. Margot hoped to continue her experiments at the Tufts laboratory of Dr. Sidney Leskowitz, and worked there briefly while finishing up her Arthritis Foundation grant. But her application for federal funding was turned down, with the indication that it might be approved if she could get more preliminary data to support the feasibility of her proposal. O'Toole, experienced in adoptive cell transfers (in which the host animal's immune cells are killed, leaving only the donor cells), was examining surface antigens that were present during bone-marrow rejection.

In mid-1985, with the help of Henry Wortis, O'Toole obtained work at the Center for Cancer Research at MIT, joining the laboratory of Dr. Thereza Imanishi-Kari. Initially the collaboration seemed promising. O'Toole needed a job; Imanishi-Kari had a project that called for a researcher with O'Toole's expertise. If O'Toole would come to MIT and work with her for a year, Imanishi-Kari promised to help her with an application to become an assistant professor of research at Tufts, where Imanishi-Kari would be moving on July 1, 1986, and where Peter worked.

It did not take long for the two women to clash. In part, Imanishi-Kari felt that O'Toole wasn't working hard enough; Thereza wasn't sure whether Margot wanted to devote the time necessary to be successful in science, or whether she wanted to stay home with her children. O'Toole, for her part, knew she wanted to be a scientist—that had been her dream since childhood; she just wasn't sure about the twenty-four-hour-a-day, seven-day-a-week model promoted by Imanishi-Kari. The tension in the lab was in marked contrast to O'Toole's happy memories of Henry Wortis's collegial lab. O'Toole readily accepted hard work as a fact (she had taken only ten days off from work after her son, Brendan, was born in 1981), but she would be happy being part of a team, rather than heading a lab of her own.

Brigette Huber, a friend of Imanishi-Kari's and a colleague of Peter Brodeur's, said one source of tension was O'Toole's involvement in a case in which she witnessed a policeman punching a restaurant worker during an arrest in Boston's Chinatown. O'Toole, who was still working in Leskowitz's lab at Tufts at the time, saw the incident, collected the names of other witnesses, made a statement to the police, and called *The Boston Globe*. Although her friends and colleagues admired her actions, Huber said O'Toole's continued involvement cut into her work at Imanishi-Kari's lab.

But Imanishi-Kari's dissatisfaction with O'Toole had a more immediate cause. In the fall of 1985, O'Toole and Imanishi-Kari's other postdoc, a Brazilian researcher named Moema Reis, had conducted some experiments with Bet-1, a rat antibody used as a reagent. These experiments seemed to confirm an important study by Imanishi-Kari on antibodies and idiotypes. Imanishi-Kari's experiments were showing that a mouse's own antibodies somehow acquired the idiotype, a special characteristic, of an injected foreign gene. O'Toole's tests were part of a study to extend Imanishi-Kari's work by determining the mechanism by which the idiotypes are copied.

But Margot was unable to duplicate her own tests. In her typically outspoken, direct way, she did not hide the discrepancies that kept coming up between her research and the work done by her boss. When her work didn't meet expectations, she asked Imanishi-Kari to show her the original scientific data; but, according to O'Toole, her boss never seemed to be able to find it.

Angry at suggestions that O'Toole was right and she was wrong, Imanishi-Kari was offended by her postdoc's audacity. Thereza would yell at Margot, telling her that she was the one making the mistake. Imanishi-Kari called O'Toole incompetent and too picky.

Her boss's criticism was debilitating to Margot, who wanted very much to be a team player. Margot initially had been excited at joining the lab, and Imanishi-Kari even planned to cite O'Toole's preliminary experimental work in an upcoming paper. Even though they seemed to be constantly arguing, O'Toole was pleased that Imanishi-Kari presented her successful early findings at an October conference.

Despite this small success, throughout the winter of 1985 and early spring of 1986 Margot stewed over her situation. She hoped to get some understanding of her troubles with Thereza from another former doctoral student who had worked in the same lab when she first arrived. Now out on his own, Charles Maplethorpe, who also had a medical degree, shared with Margot his own feelings of uncertainty about the academic world in general, and Thereza in particular. He thought Imanishi-Kari acted in an unusually secretive manner in the lab. And he was also suspicious of an encounter he witnessed at the lab during which, he said, Nobel laureate David Baltimore's postdoc David Weaver raised questions with Imanishi-Kari about the Bet-1 reagent. After talking with Maplethorpe, O'Toole began to feel that her career was at a crisis point.

■

Finally, Margot acknowledged failure. But not her failure. She had come to believe that Thereza's conclusions must be wrong. In fact, in March she believed she discovered an error by Moema, who had since returned to Brazil, that invalidated their initial confirming results.

Imanishi-Kari, too, had had enough. She pulled O'Toole off the project and ordered her to take charge of mouse husbandry. Although breeding the mice actually required some sophisticated knowledge, the job certainly was not the launching pad for her career that O'Toole had envisioned when she took the job with Imanishi-Kari. O'Toole complained to Dr. Herman Eisen, head of the MIT immunology department, who had recruited Imanishi-Kari; to Dr. Mary Rowe, MIT's ombudsman and assistant to the MIT president; and to a number of her colleagues. No one came to her aid. She also tried to approach Henry Wortis, but he did not want to get involved in what he viewed as a personality conflict.

When Imanishi-Kari told O'Toole to prepare a paper on a study that O'Toole had done, which was closely related to Imanishi-Kari's work, Margot refused. O'Toole claimed that she was instructed several times to simply omit some of the facts in order to bolster the conclusion. Believing that if she did as she was told, the paper would be misleading, O'Toole simply stopped arguing and did nothing.

Fed up, O'Toole gave notice that she was quitting the lab, but she agreed to continue caring for the mice through May 1986.

Although she had not made any headway with Imanishi-Kari, O'Toole was surprised when one day in early April her boss admitted to her that some of her actual data agreed with O'Toole's results. This was a particularly surprising admission, since Imanishi-Kari's interpretation of the results was to be published soon in a paper that O'Toole had critiqued. Nonetheless, the paper on Imanishi-Kari's experiments was published shortly afterward, in the April 25, 1986, edition of the noted scientific journal *Cell*. Among the acknowledgments was the following sentence: "We are indebted to Henry Wortis, Margaret [*sic*] O'Toole, Martina Boersch-Supan, and Elaine Dzierzak for critical reading of the manuscript."

Margot also learned at a chance meeting with Brigette Huber, an associate professor of pathology at Tufts and once a member of her thesis advisory committee, that Thereza was characterizing the dispute as a "personality conflict." O'Toole fumed. The problem was the science: Imanishi-Kari's conclusions were unsound, she told Huber. Brigette, a friend of both women, agreed to look at the data and see if she could help. She would be less busy in a couple of weeks; Margot should visit then.

■

On May 7, 1986, just about a year after her arrival at Imanishi-Kari's lab, Margot pulled a three-ring binder off a research shelf to check on the pedigree of a mouse. The binder contained the breeding records Moema Reis had left behind upon returning to Brazil. Reis had worked on the disputed experiments and was listed as one of the six authors of the *Cell* paper.

Flipping through the binder, O'Toole found seventeen pages that immediately struck her as being out of place. The pages of numbers and Greek letters would have been unintelligible to a layperson. Only to a scientist did they have any meaning. To O'Toole, they meant vindication.

What she had in her hands was some of the raw data on which Imanishi-Kari had based her conclusions. Leaning against the lab bench, O'Toole turned the pages back and forth, rereading them. It was clear to her that the research did not support the conclusions in the published paper.

She duplicated the pages and examined them during her lunch break. Adrenaline rushed to her head. Margot felt an emotional surge akin to vertigo.

"It's not me!" she called out to no one in particular. The universe again made sense to her.

■

Her joy was short-lived. The scientific dispute her discovery set off would precipitate a nationally publicized academic brawl that had disputants accusing each other of either sanctioning fraud or practicing scientific McCarthyism. Over the next few years, O'Toole would find her scientific career in shreds; Imanishi-Kari's reputation would be in ruins; even the president of Rockefeller University, Nobel Prize–winner David Baltimore, who had put his personal stamp of approval on the publication of the study, would be dragged into the controversy and eventually forced to resign. A congressional committee would very publicly investigate the case—and get shellacked in turn by scientists around the nation. National Institutes of Health procedures for rooting out fraud would be altered and tightened—and still criticized. And the hundreds of thousands who suffered from the diseases that Thereza Imanishi-Kari's work might have led to a cure would have to look elsewhere for hope.

But in May 1986, O'Toole did not have a clue as to the forces that she had unleashed. She again asked Imanishi-Kari for the records to her experiments. Again, the scientist did not produce them. Margot told her husband of the data that she had found, and they both agreed that she would do what she felt was necessary. The O'Tooles, after all, had a "history of doing the right thing." Peter suggested that Margot show Huber the seventeen pages. After that discussion, Margot tried to insulate her husband from her actions on the *Cell* paper and did not talk to him about them. On May 9, she took Huber up on the offer of help and gave her a copy of the seventeen pages along with her reasons for doubting the *Cell* paper.

Huber quickly realized that O'Toole's questions could balloon into a major controversy. She did not want the dispute "on my shoulders," but she also knew that Margot, as a lowly postdoc, could not deal with it alone. For advice Huber called a scientist with more expertise, Dr. Robert Woodland, an associate professor of molecular genetics and microbiology at the University of Massachusetts Medical School. Woodland, who considered himself a friend of both Imanishi-Kari and O'Toole, suggested going to Dr. Wortis, whose immunology division at Tufts was planning to hire

Imanishi-Kari. Margot wasn't sure: Henry had close personal and professional ties to both Thereza and herself. But Brigitte pushed her: Wortis should know of the problems because he was in the process of applying for a grant as a co-investigator with Imanishi-Kari to continue this work. Presumably, the Tufts faculty would be concerned about the quality of a prospective colleague's work.

O'Toole also tracked down Martin Flax at home. Flax was the dean of the Tufts pathology department, which includes the immunology division. On the phone, O'Toole informed him of the events so far. Flax encouraged her to bring the matter to MIT authorities: MIT was where the work had been done, and MIT was responsible. When O'Toole said that Wortis would be reviewing her complaints, Flax told her to proceed.

Wortis had tried previously to stay out of the middle of the growing dispute between O'Toole and Imanishi-Kari. "Dr. O'Toole was a bright, insightful scientist who had a good understanding of cellular immunology. . . . I recommended O'Toole very highly," he later recalled. "Many months later, I learned they were having problems working together. O'Toole expressed strong concerns over the lifestyle that was being imposed on her and Imanishi-Kari was concerned over O'Toole's willingness to put in the necessary time. I did not want to get involved. I told O'Toole to work it out."

Now Wortis had no choice but to get involved. O'Toole asked him, he contended, to investigate whether the mouse used as a control actually was a "true normal"—that is, one that did not have genetic peculiarities that might skew the results. Only a true normal could be used as a comparison to a mouse injected with the foreign gene. But the seventeen pages indicated that the cells from control mice had a relatively high incidence of the foreign gene. Further, she wanted Wortis to investigate whether the Bet-1 reagent was a good enough indicator to detect the foreign mouse gene. If Bet-1 did not react with a cell, then the gene present was actually one of the mouse's own making, and its presence could be evidence for the paper's conclusion. No one was asserting fraud at that time, and Wortis looked for none.

Without identifying O'Toole as the person who had raised questions, Wortis was supposed to arrange to meet with Imanishi-Kari on the morning of May 15. Shortly after the meeting was scheduled to begin, according to O'Toole's distinct recollection, Imanishi-Kari called her, yelling over the phone and accusing her of vindictiveness. Thereza's anger was such that she threatened to sue, Margot recalled. O'Toole tried to calm her, stressing that it was Imanishi-Kari's friend and supporter Henry Wortis who was looking into the questions. (Imanishi-Kari did not remember such a phone call; she said threatening to sue would be out of character.)

Wortis, Huber, and Woodland met Imanishi-Kari on the evening of May 16 at her MIT lab and went over records and data. The members of the ad hoc panel agreed that even if there was an overstatement of Bet-1's capabilities—in other words, the reagent was reacting to both the foreign and endogenous antibodies—that would have actually strengthened the conclusions of the *Cell* paper. This was because if the reagent did not discriminate between foreign and endogenous genes but reacted to both, the scientists would have been overestimating the amount of transgene product and underestimating the number of endogenous genes. Some idiotype-positive endogenous genes would have been counted as the foreign genes.

The informal panel said, however, that the *Cell* paper's results did not depend on what Bet-1 does or does not do.

As for the control mouse, Imanishi-Kari told them that the mouse marked as a normal actually was a transgenic mouse that had been mistyped. Another control mouse was used instead, she said.

"At the conclusion of the meeting, we all agreed that there were no significant problems with the paper," Wortis recalled.

Huber called O'Toole the next day. She described the data they had seen and said there was no discrepancy from the published data. But O'Toole insisted that what they had seen was not adequate to resolve the problems with the paper.

Although Wortis was surprised when Huber informed him later that his former doctoral student was still dissatisfied, he agreed to a second meeting at Imanishi-Kari's lab on May 23. "The meeting was intense and at times heated," Wortis later told a congressional panel. "However, it was not unlike other scientific meetings and discussions I have witnessed and taken part in."

O'Toole stood up at the end and offered her hand to Imanishi-Kari. Although Imanishi-Kari refused to shake hands with O'Toole, Wortis believed that they had agreed to disagree on the conclusions of the study.

No report of the meetings was written up until the following year, when one was produced in response to a federal inquiry.

However, a memo of the ad hoc panel that Wortis wrote up in 1987 for Henry Banks, dean of Tufts Medical School, concluded in capital letters: "NO EVIDENCE OF DELIBERATE FALSIFICATION. NO EVIDENCE OF DELIBERATE MISREPRESENTATION. ALTERNATIVE INTERPRETATIONS OF THE EXISTING DATA CAN BE MADE, BUT THAT IS THE STUFF OF SCIENCE."

But the memories of the participants varied vastly. Woodland, who had ruptured his Achilles tendon in a volleyball accident, was not there for the second meeting, but O'Toole attended. Imanishi-Kari presented two pages of data and said she resented having to take a week to generate the data to satisfy O'Toole's objections. O'Toole asked to see records of the experimental steps that would have had to precede the results that Imanishi-Kari brought to the meeting.

"I want to see the original data," O'Toole said. Imanishi-Kari was silent.

Deal with the data at hand, Wortis told O'Toole.

According to O'Toole's memory of the meeting, Imanishi-Kari confirmed that she had not done a series of isotyping experiments described in the paper. Furthermore, O'Toole recalled that Wortis and Huber acknowledged that the problems with Bet-1 were serious. Imanishi-Kari "said all the problems were the results of inadvertent errors and I did not question her explanation," O'Toole said later.

Believing that her assertions—that the data did not support the published claims—were confirmed, O'Toole left the meeting relieved. She expected that the paper would be retracted.

Much to her surprise, nothing was going to be done. As far as Wortis was concerned, the matter was reviewed and closed.

In a letter written a year later to Tufts Medical Dean Henry Banks, Wortis reported that O'Toole's work "did not include any attempts to confirm the origination observations—rather, she was testing a new and highly conjectural hypothesis. The experiments failed and this became a matter of contention."

Nonetheless, Wortis said that he "felt that on the face of the evidence at hand, an inquiry was necessary." He wrote that his ad hoc group met with Imanishi-Kari and went over the original notebooks. He told Banks that he had made no written report at the time, but did speak to Martin Flax, the Tufts dean of pathology, to tell him the results of the meeting.

Flax also asked Dr. Sidney Leskowitz, Tufts's senior immunologist, in whose lab O'Toole worked briefly before joining Imanishi-Kari's group, to look into her complaints from a different angle: the hiring of Imanishi-Kari. After talking to her superiors, colleagues, and students, Leskowitz reported that Imanishi-Kari was considered a first-rate scientist, a hard worker, and a helpful colleague, Flax later said. There was concern about her relations with students, but Flax said that might "have been due to the poor quality of students assigned her." (Leskowitz died in 1991.)

The immunology group at the pathology department discussed Leskowitz's report and voted to recommend that Imanishi-Kari be appointed an assistant professor. The vote was unanimous; even O'Toole's husband voted for Imanishi-Kari, in part because he believed that simple fairness dictated a yes vote, since the job had already been offered and there had been no substantive investigation to warrant leaving her without a job. (Imanishi-Kari had already cut her ties to MIT by that time.) And Brodeur assumed that Wortis had reached some understanding with Margot; she had not discussed the case with her husband. (Three years later, congressional investigators said they were told by scientists at Tufts that even O'Toole's husband did not have confidence in his wife's understanding of the science, citing his vote for Imanishi-Kari. When Margot later told Peter about the comments, he was shocked that Tufts scientists had used his efforts to be fair and neutral against his wife.)

Brodeur also tried to be a good colleague: He turned over his lab to Imanishi-Kari while her own was being renovated and he made do with less suitable quarters. Brodeur said he managed to work with people at such odds with his wife by "compartmentalizing" and keeping relations on a professional basis. Huber and others said he was indeed a good colleague and simply agreed not to speak with him about the controversy.

■

In the meantime, O'Toole had kept Mary Rowe, the MIT president's assistant and ombudswoman, up-to-date about the Wortis review. Rowe had pressed O'Toole to bring charges and go through official MIT channels. O'Toole refused; she only wanted a retraction of the paper if her suspicions proved correct. Formal charges, she believed, would destroy any possibility of a collegial resolution of the scientific dispute.

While she remained opposed to filing formal charges, a conversation with Huber changed O'Toole's mind about pursuing her complaints at MIT. O'Toole says Huber told her to drop her complaint or "there will be less money for everyone in the department, especially for people just starting out." Huber denied the conversation. But O'Toole insisted that she heard what she believed was a threat against her husband: "Peter was just starting out."

"They tried to turn my love and marriage against me," O'Toole said. The purported threat was pivotal; it convinced her to pursue her complaint at MIT, which she did without telling Brodeur. "I had to do what I did. I had to go forward based on trusting that Peter was who he was. They held him hostage."

O'Toole went to Rowe and admitted that her decision to go the informal Tufts route had not had satisfactory results. Rowe insisted that O'Toole talk to Gene Brown, dean of MIT's School of Science. Rowe told her, Margot said, that rumors that she had falsely accused Imanishi-Kari were spreading through the labs. Rowe later declined to talk about these events to federal investigators on the grounds of confidentiality, even though O'Toole waived any privilege in the matter.

O'Toole met with the dean on May 29. She remembered that Brown told her to either bring charges of fraud or drop the challenge, and she chose the latter option. Brown, however, recalled a collegial conversation in which he said he would ask Herman Eisen, MIT's senior immunologist, to look into her concerns. Eisen was well respected in the scientific community, and known as a friendly schmoozer.

Indeed, the next day, Eisen called O'Toole and arranged to meet her that afternoon in Woods Hole, where he had a residence, to discuss the dispute. She brought copies of the seventeen pages found in Moema Reis's mouse binder and explained her concerns. What she had to say was not easily understood, Eisen later said; he thought Margot was overwrought, so he asked her to put it in writing. He wondered aloud if she was charging fraud, but O'Toole insisted that she was concerned only with error. O'Toole wrote a memo, dated June 6, just two months after the publication of the *Cell* paper.

The memo was carefully bland, purposefully so on the advice of Mary Rowe. There were no charges of fraud or misconduct. "In response to requests from you and Dr. Mary Rowe, I have prepared this report on what I perceive as serious weaknesses in the data presented in *Cell* . . . by Weaver et al.," was O'Toole's introduction to the memo. She concluded: "I would be interested in your comments."

Her four basic concerns remained:

1. The Bet-1 reagent used for detecting the foreign antibody also detected the mouse's own antibody. This lack of "specificity" would have made it difficult to know if the experiment had been a success.

2. The number of control mouse antibodies that tested positive for the idiotype was actually significantly higher than those shown in Table 2 of the *Cell* paper. Also, the experiment on the control and those on the transgenic cell lines shown in Table 2 were done at different times. This could possibly affect the outcome.

3. An inadequate test was used to show that the foreign antibody was not made in the majority of idiotypic-positive cell lines. The available evidence, O'Toole said, indicated that most of the cell lines made the foreign protein. Or, in other words, the foreign gene had apparently "turned on" once it was injected, and had begun making the antibodies itself rather than spurring the mouse's own genes to create the copycat antibodies.

4. Finally, the anti-idiotypic radioimmunoassays were more sensitive to transgene expression than were the mRNA-detecting assays. This meant the mRNA assays might not have been picking up low-level expression of the transgene that might have been detected by the radioimmunoassays, and some cells would be inaccurately scored as non-expressors of the foreign gene.

Basically, O'Toole was saying that according to the data she found in the seventeen pages from the binder, most of the idiotype-positive antibodies came from the inserted foreign gene.

But she actually was saying much more than that. On page 2 of the memo, O'Toole noted that the cells in Table 2 "were not checked for isotypes other than mu, according to Dr. T. Imanishi-Kari and Dr. M. Reis; the statement on page 250 that the majority of these hybridomas express gamma2-b is based on an analysis of a number of hybridomas from another fusion." Eisen would later say that he was not told that experiments had not been done. But the meaning of O'Toole's memo was just that: Tests key to the central conclusion of the *Cell* paper had not been done as published.

Shortly after receiving the memo, Eisen was back on the phone to O'Toole. How could anyone explain how the misrepresentations could have been published without suggesting fraud as the cause?

O'Toole, again, explained that she was only interested in the accuracy of the science; judging fraud was not her concern.

■

On June 16, Eisen called Imanishi-Kari, David Weaver, David Baltimore—the principal co-authors—and O'Toole to a meeting at Baltimore's comfortable office at the Whitehead Institute for Biomedical Research, the new research facility that bordered the MIT campus. Baltimore, a Nobel laureate, was the institute's first director.

Again, memories differ. According to O'Toole, Imanishi-Kari said she had not brought the critical records along because she did not think she needed them, and besides, they were written in her native Portuguese. Baltimore assured her that he did not need to see other data. They looked at graphs that Imanishi-Kari said she had generated in Germany some years previously, and they reviewed some of Weaver's records.

O'Toole recalled that after they examined the records for Table 2, Baltimore said, "You can't tell anything at all, one way or the other, from this." She also remembered Imanishi-Kari confirming that she had misrepresented the Bet-1 data in Figure 1 of the paper. When questioned why she had done this, Imanishi-Kari replied that Moema must have "got that result once." Baltimore said he would discuss the issue further with Imanishi-Kari in private.

"Now, you should drop this for your own good," Baltimore, the senior scientist, told O'Toole. "A lot of my time goes into calming people like you down."

These challenges occur often because junior people don't see all of the experimental records, he explained. But O'Toole remained insistent: She had legitimate access to the records in question, she said. Then all she could do now, Baltimore responded, was to write up her challenge and send it to a scientific journal, but in that case he would personally write a rebuttal. The implication was clear to O'Toole: Who would believe her over a Nobel laureate?

The other participants maintained that Imanishi-Kari did not admit to misrepresenting data. Baltimore said he suggested in good faith that O'Toole submit her comments to a journal; he was not trying to intimidate her out of doing so.

When O'Toole told Baltimore that she planned to drop the matter, now that it had been reviewed at MIT and Tufts, he responded, "Oh well, then there is no problem." At least part of the study held up, so no correction was warranted, he said.

Despite her stated willingness to let the matter drop, O'Toole continued to argue with Eisen after the meeting. She couldn't understand why, given Imanishi-Kari's acknowledged misrepresentation, he did not examine the necessary records and was not pushing for the paper to be retracted or corrected.

But Eisen cautioned O'Toole that if she continued to pursue the challenge, it "would seem to indicate vindictiveness," according to her recollection.

The next day, Eisen wrote up his view of the meeting and O'Toole's challenges in a memo to Gene Brown, dean of the school of science; Maury Fox, chairman of the biology department; Mary Rowe; and Phillip Sharp, director of the MIT Center for Cancer Research. After a quick review of the issues, Eisen noted that "without more extensive study it is not clear whether the paper's conclusion[s] or O'Toole's are correct; and more extensive study will doubtless be undertaken by Imanishi-Kari and others.

"I do not think that I or anyone else present at the meeting felt that Margot O'Toole's disagreements were frivolous," he continued. "They are indeed based on pretty carefully thought out ideas of the limitations of the analytical methods." Nonetheless, he wrote that such disagreements are "not uncommon in science, and they are certainly plentiful in immunology. The way they are resolved traditionally, and effectively, is by publishing the results and having other laboratories try to repeat and evaluate them. . . . This is the way science operates; and in fact it is the kind of contentiousness seen in this dispute that helps drive the science 'engine.' Therefore, it appears to me that the entire exercise is not an unusual one except for the intensity of the feelings generated and the circumstances concerning Dr. Imanishi-Kari's pending appointment at Tufts."

Despite his support of O'Toole's analysis at the beginning of the memo, Eisen both criticized and praised the way she had challenged the paper. "Before closing, I should call attention to the extra burden of pain and discomfort created by the way in which

Margot O'Toole's claims have been aired,'' Eisen wrote. Baltimore, he said, did not hear of her disagreement until many people at Tufts and MIT had already heard of it and were discussing it. "On Margot O'Toole's behalf, however, I would also add that it took a rather considerable amount of courage to face a senior scientist whose scientific judgment she was questioning in a serious way."

There was no reference to fraud, misconduct, or misrepresentation.

For some reason, however, Eisen filed his memo and did not send it to anyone. Another memo dated December 30, 1986, some months later, was sent to Maury Fox and others, apparently because of inquiries from NIH. This time, Eisen referred to "the allegations of misrepresentation" that were brought by O'Toole. Three of O'Toole's questions seemed to be based on matters of judgment, Eisen wrote, and not on "evidence of misconduct."

However, O'Toole's allegation that the Bet-1 reagent was not specific "was disturbing because it raised a serious question about deliberate misrepresentation of data." But he wrote that Imanishi-Kari explained that the reagent does detect the transgene to a much greater extent than it does the endogenous gene. Eisen discussed the problem with the NIH scientist whose lab developed the Bet-1 antibody and who, Eisen said, agreed with Imanishi-Kari's claim. (Some scientists, however, note that Imanishi-Kari used Bet-1 under different circumstances.)

"My conclusion is that O'Toole is correct in claiming that there is an error in the paper; but it is not a flagrant error," Eisen wrote. "The correction would be too minor to rate a letter to the journal; it certainly does not warrant a retraction, especially because the paper contains a substantial body of other data that is clear and impressive."

The other issues raised by O'Toole were "largely matters of interpretation and judgment [and] are best dealt with by allowing the scientific process to take its course. Other laboratories are trying to extend the findings. In this way we will know if the interpretations are right or wrong."

In testimony before the congressional panel three years later, David Baltimore described his role at the meeting as that of a bystander who asked occasional questions.

"I left that meeting convinced that there was no serious error in the *Cell* paper," he said. He was impressed by O'Toole's analysis and "believed that further experimentation might validate certain of her points and that therefore certain of her criticisms could be a valuable contribution to science."

However, O'Toole was anything but satisfied with the outcome of the meeting. She complained to Mary Rowe about how Eisen had handled the dispute, but she got little sympathy there. Rowe said Eisen had not reported any misrepresentation to her. Indeed, Eisen would not mention anything about misrepresentation, at least officially, until his December memo. The matter, O'Toole remembers Rowe saying to her, is now "in the hands of God."

Margot finally decided she had gone as far as she could with her concerns. The turmoil had taken a personal toll on her family; she was concerned about any backlash hurting her husband's career at Tufts. Although Peter thought that she could reestablish her science career by working in his lab at Tufts, Wortis warned him that it would not be a good idea. Some of O'Toole's friends wondered how her husband could continue working in the same department as Imanishi-Kari and her supporters. But it was very clear to O'Toole and Brodeur. "We decided to protect the one job we had between us, and I suggested that he offer the position to someone else," O'Toole recalled. "I have a great husband. Peter is a real rock. He knows right from wrong."

They were unable to afford the mortgage on the larger house they had bought for their growing family (Dylan was born in 1987, and Paul in 1990), and so they leased it out and moved in with Mrs. O'Toole. (Margot would later be angry when a reporter wrote that she had lost her home and was forced to move in with her mother; she took that as insulting to her mother, who had been wonderful to them.)

Mrs. O'Toole gave her daughter a copy of a Yeats poem, "To a Friend Whose Work Has Come to Nothing."

Now all the truth is out,
Be secret and take defeat
From any brazen throat,
For how can you compete,
Being honor bred, with one
Who, were it proved he lies,
Were neither shamed in his own
Nor in his neighbor's eyes?

Bred to a harder thing
Than triumph, turn away
And like a laughing string;
Whereon mad fingers play
Amid a place of stone,
Be secret and exult,
Because of all things known
That is most difficult.

Mrs. O'Toole thought the poem would help give her daughter the courage to accept what had happened and to realize that it was not her fault. She also hoped that it might help her continue her challenge, help her to right a wrong. The mother had thought the daughter had been a little naïve not to have realized in the beginning what would happen when she raised her questions.

But Margot did not want to continue her fight. The poem, which she had first read as a child in Dublin, was a source of great comfort, but what she took from it was not that she should continue her challenge but that having done her duty to try to correct the paper, she did not have to become part of what she saw as corruption. What O'Toole had not counted on was Charles Maplethorpe, the former doctoral student in Imanishi-Kari's lab.

Despite his earlier warning that she should just remain quiet and get her own work done, Maplethorpe had decided on his own to keep the issue alive. When Maplethorpe heard about the plans for the Wortis meeting, he began to worry for Margot. He feared she would get run over by the senior scientists she was challenging; it would be good to get an impartial observer involved, if only to know that the review was taking place. But who would be interested in getting involved in the dispute? Two engineering friends, who could never quite believe Maplethorpe's stories of political intrigue in the laboratories, told him about a *New York Times* article they had read recently about the work of two NIH scientists who had rooted out scientific misconduct at Harvard and Emory despite threats of lawsuits and difficulties in getting their report published.

Those two scientists at the National Institutes of Health, Maplethorpe thought, just might be interested in O'Toole's case as another study in scientific misconduct. But he did not tell O'Toole about them quite yet.

TWO

The Experiment

■

The dispute over what really happened in Thereza Imanishi-Kari's laboratory would engulf the lives and threaten the careers of three dedicated scientists; pit the scientific community against a hard-nosed and relentless congressman; involve charges of fraud, sexism, government neglect, overzealous prosecution, blackballing, and incompetence at the National Institutes of Health; produce headlines and self-righteous editorials in the national press for years; polarize the scientific community; and profoundly shake the public's confidence in the disinterest and probity of scientists. Yet at the heart of this complex and wrenching affair were two tiny mice, one white, one black.

Both mice were of common strains used by scientists. The strains had been deliberately inbred so intensely so that they had lost their genetic diversity. Each mouse was genetically identical to its siblings and cousins in its own particular strain. These two strains of mice were known by the scientifically dispassionate names C57BL/6 and BALB/c.

There were, however, important differences between the antibodies of these two mice. Those differences, characterized by the Japanese-Brazilian immunologist Dr. Thereza Imanishi-Kari and other scientists in the late 1970s and early 1980s, made the two mice attractive to one of the United States' most prominent scientists, Dr. David Baltimore. His collaboration with Imanishi-Kari on the *Cell* paper was a key factor in the attention the controversy would receive; his stature as a Nobel laureate made the story good copy; and his arrogant response to O'Toole's concerns fueled passions on both sides of the dispute. If the characteristics of two mice were the cause of the "Baltimore case," human frailty multiplied the consequences.

■

Most adults—and schoolchildren, too—believe they have a working knowledge of their bodies' immune system: Some bit of bacterium gets under the skin and the body throws a hail of disease-fighting things called antibodies against it. Immunology is a lot more complicated than that to scientists, who map the chains of amino acids, regions, and microcomponents of antibodies in order to get a precise explanation of how the immune system works. If they know how it works, they can learn how to control it. But immunology is an arcane, ambiguous science, in which it's often hard to grab hold of substance and definitions are frequently circular.

Antibodies are proteins that recognize microorganisms that invade the body and can make a person ill. Antibodies are sometimes described as "guided missiles," but analogies to snowflakes and dust-collectors might be more apt.

Antibodies resemble snowflakes in that no two are alike. Each antibody is produced by the body to go after a particular antigen. An antigen—here's the circular part—is a substance that provokes the production of an antibody specifically for that substance. Now the dust-collector notion: The antibody's job is to collect the bit of antigen by grabbing hold of it—that is, by binding to the foreign substance. The immune system then has other weapons that try to kill the attackers.

Scientists have been baffled by how the body knows how to produce a veritable snowstorm of antibodies, each one unique, as they are needed.

A clue may be found in a characteristic of the antibody: the idiotype. The idiotype is a determinant of the section of the antibody structure associated with the antibody's capability to bind with antigens.

In 1974, Dr. Niels K. Jerne, a scientist born in London of Danish parents, theorized that the body regulates the components of its immune system to fight off disease through cell-cell interactions, a concept known as the network theory. According to Jerne's theory, the idiotype acts like an antigen and stimulates B, or white, cells to produce other antibodies with anti-idiotypes: those antibodies in turn provoke the production of a third set of antibodies with idiotypes, and so on. Every antibody would then be linked to other antibodies in a network. Idiotypes are also found on B cell surface antibodies, and thus the anti-idiotypes could regulate the white cells producing their target antibodies. The body's immune system might therefore be regulated by an idiotype network.

In 1984, at the age of seventy-two, Jerne won a Nobel Prize for his radical 1974 work, as well as for his explanation of the way antibodies are generated to attack viruses or other invaders of the body. "This view on the nature of the immune system constitutes the basis for modern immunology," said authorities at the Karolinska Institute in Stockholm in announcing the award, which Jerne shared with two other immunologists. (At a celebration afterward, Jerne quipped, "It would have been nice if it had come earlier to convince my brothers and sisters that I am not the oddball they regarded me for a long time. I will enjoy the prize and enjoy life.")

The *Cell* paper by Imanishi-Kari, Weaver, and Baltimore clearly caught the scientific community's attention. The authors announced results that suggested evidence of Jerne's networking theory, although they did not know the mechanism by which the networking would operate.

Imanishi-Kari, Baltimore, and their colleagues used the C57BL/6 and the BALB/c mice for their experiments—the C57BL/6 for its known immune system and the BALB/c mouse for an antibody gene that was unlike those in the black mouse.

In order to determine what the black mouse's immune system was doing, the scientists had to identify the components of the antibodies that were being produced. The paper dealt with antibodies, also known as immunoglobulins (Ig), found in the gamma globulin part of serum, produced by B cells. The antibody is made up of chains of proteins, or amino acids. Each Ig is made up of heavy chains and light chains.

Two classes of the Ig heavy chain are mu and gamma, found in two of the five types of immunoglobulin, IgM and IgG. IgM makes up about 10 percent of all immunoglobulin in normal human serum (the watery part of the blood) and is the largest and most complex of the Ig molecules; while IgG makes up about 75 percent of the immunoglobulin.

The mu and gamma heavy chains, which can be recognized by other antibodies, are said to be of different isotypes. No one antibody molecule can have both a mu and a gamma chain. But an IgM molecule can have chains with different allotypes of mu: mu^a and mu^b are said to be different allotypes, or variants of the isotype. The allotypes, which can be distinguished from each other with proper reagents, are found in the areas known as the "constant regions" of the heavy and light chains.

Now there is still another creature of immunology that is key to the controversy that engulfed Baltimore and Imanishi-Kari: the idiotype, a signature of the molecule's "variable region." The variable region is the area of the antibody's structure that determines what the antibody will bind to. Two molecules of different isotype or allotype may have the same idiotype.

In the early 1980s, Baltimore wanted to take advantage of a new technique—the injection of foreign genes into mouse embryos to produce "transgenic" mice—to explore the production and control of antibodies. At the time, there were not many scientists schooled in the technique of isolating a mouse's fertilized egg, injecting DNA into the egg's nucleus, and then reimplanting the egg into a female mouse. So Baltimore began a collaboration with Dr. Frank Costantini, a developmental biologist at Columbia University's department of human genetics and development, who could create the transgenic mice necessary for Baltimore's experiments. They used a mu immunoglobulin heavy-chain gene to examine whether it would have any effect on the endogenous genes of the mouse under study, according to Baltimore.

"Our motivation was the basic understanding of how things work," said Costantini, who had never collaborated with Baltimore before.

Costantini used a gene that had the heavy chain of a particular BALB/c white mouse idiotype, known as 17.2.25. Using the microinjection technique, he inserted the gene—now known as a transgene—into the embryo of a C57BL/6 black mouse, which did not normally have that particular idiotype. They had gotten the 17.2.25 idiotype from Imanishi-Kari, who had identified the idiotype in her earlier work.

As Baltimore would later tell the scientific community, "There was a long-range possibility that by putting a new idiotype into the C57BL background, there might be interesting immunologic consequences." The experiments showed an increased production of antibody, or "expression," of the transgene. Indeed, Baltimore said: "Studies of the expression of the inserted 'transgene' provided new insights into the regulation of antibody production."

The experiments whetted his interest in what the transgene, the one taken from the BALB/c mouse, was actually doing to the C57BL/6 mouse's own antibody genes. But Baltimore felt that while his lab could do the molecular biology and examine the cells' RNA, which carries the genetic messages ordering the production of antibodies, it did not have the expertise to analyze the antibodies produced by the transgenic mouse. In 1984, he brought in Imanishi-Kari and her lab to do the immunology work. Costantini, once again, bred and screened the mice and sent them up to MIT for the experiments. Imanishi-Kari used the Costantini mice for some of the experiments and for breeding of other mice for additional tests.

The astonishing results were written up in a paper published April 25, 1986, in *Cell,* under the long, awkward title "Altered Repertoire of Endogenous Immunoglobulin Gene Expression in Transgenic Mice Containing a Rearranged Mu Heavy Chain Gene." The authors, as listed in order, were David Weaver, Baltimore's postdoctoral student; Moema H. Reis, Imanishi-Kari's postdoc; Christopher Albanese, Imanishi-Kari's technician; Costantini; Baltimore; and Imanishi-Kari. The paper is often referred to by scientists in shorthand as "Weaver et al.," after the lead author.

The two teams announced that they had found significant evidence suggesting that the mouse had produced its own antibodies that had the idiotype of the transgene. Somehow the mouse's antibodies had acquired the transgene's idiotype, the characteristic associated with grabbing hold of specific antigens. Although the transgene was present, it had not "turned on"—it was not active—and yet something had happened. According to the paper's summary: "The expression of endogenous genes mimicking the idiotype of the transgene suggests that a rearranged gene introduced into the germ line can activate powerful cellular regulatory influences."

They had found evidence of "idiotypic mimicry" and support for Jerne's networking theory. And although the authors did not state it directly, others saw hints that the results could lead to the discovery of how to harness a person's own antibody network to fight such immune-system disorders as AIDS and lupus. Indeed,

Imanishi-Kari would later reflect on the possible impact of the paper on lupus research.

The paper told of how Imanishi-Kari's team used two types of antibody reagents—anti-allotype and anti-idiotype—to distinguish between the transgene and the mouse's endogenous genes. Two anti-allotype reagents were used to distinguish between the mu constant regions of the transgene and the endogenous genes; the BALB/c mouse's mu constant region has a determinant not present in the other strain. The scientists reported that they found that endogenous Ig "very frequently mimics the idiotype, but not generally the polypeptide structure of the transgene product."

The transgene, an M-class immunoglobulin, or IgM, was of type mu^{a} and idiotype 17.2.25. Before the injection, the mouse's own antibodies were supposed to be allotype mu^{b}. Imanishi-Kari's lab used two other iodinated, or radioactive, antibodies as reagents (substances used to detect other particular substances by the chemical reactions they cause) to determine what happened to the mouse's immune system.

The two reagents were monoclonal rat antibodies; monoclonal antibodies, an important research tool, are single cells formed by the fusion of cancer cells and B cells, which then flourish and can be easily shared with other scientists for their use. One reagent was AF6, which binds with mu^{b}, the allotype of the endogenous genes, but not with the mu^{a} allotype of the inserted transgene. According to the paper, a second reagent, Bet-1, distinguishes, or is specific to, the mu^{a} of the transgene and not the mu^{b} of the mouse's own gene. The authors were quite definite about that: Figure 1 of the paper shows data demonstrating such specificity, and in the text the authors stated it in terms that would be understandable to other scientists: "The C57BL/6 anti-N IgM antibody, P9.37.9, bound only to the anti-mu^{b} allotype (Figure 1)."

A third antibody reagent, known as anti-idiotype, would detect the idiotype of the transgene. This reagent was a polyclonal guinea-pig antibody; because it was polyclonal and did not grow as vigorously as the fused cancer cells, its supply was limited.

The scientists removed B, or white, cells from the lymph nodes and spleen of the engineered mice, fused each with a cancer cell, and created "hybridomas," or cell lines that would flourish in a culture. The hybridomas and the supernatant—the liquid that surrounded them—produced antibodies and were routine research tools.

The antibodies from the transgenic mouse were poured on a plastic plate of wells—modernized test tubes—first coated with the anti-idiotype reagent. Those antibodies that stuck to the plate were those that had the idiotype: They were said to be idiotype-positive. Then those antibodies were passed over a plate coated with the Bet-1 to show if any were of the mu^a allotype. Thus, the only antibodies left were those that were mu^a and had the transgene's idiotype, 17.2.25. The same process was repeated, but using the AF6 reagent to look for endogenous mu^b.

Figure 1 purportedly showed that the AF6, indeed, bound only to the mu^b antibodies and Bet-1 bound to the antibody of the mu^a transgene, with an insignificant amount of reaction to the endogenous antibodies; the text stated that there was no cross-binding. The paper stated that according to the data represented in Figure 1, "a significant level of mu^b allotype [the endogenous gene] was also observed to have the 17.2.25 idiotype."

The scientists then took 340 spleen and lymph hybridomas from a transgenic mouse, as well as 244 from a control mouse, to look more carefully at the allotype and then the isotype. As was pointed out earlier, single antibody molecules cannot have proteins of different isotypes. If the molecules were found to be of an isotype different from that of the transgene, then the scientists would have evidence that the endogenous genes were mimicking the foreign idiotype.

The paper states that the authors analyzed the hybridomas—cell lines—from the mouse lymph nodes and spleen for the 17.2.25 idiotype, the heavy-chain allotype, and the isotype. According to the author's Table 2, about 28 percent of the transgenic hybridomas from the spleen and 68 percent from the lymph nodes produced Ig with the 17.2.25 idiotype. Less than 1 percent of hybridomas from the control mice, those that did not have the foreign gene inserted, reacted to the reagent; this finding indicated next to no presence of the transgene idiotype, as would be expected.

Of the 172 spleen and lymph node hybridomas that were idiotype-positive, only 53 produced IgM. The 119 remaining clones, or single cells, produced other Ig isotypes, "the majority being gamma2B," the paper stated. The authors noted, however, that in the published paper they were not showing the data that supported the claim. Table 2 shows that only one of the 244 control cells was positive for the 17.2.25 idiotype.

This finding of the G class of immunoglobulin, or IgG, in the majority of hybridomas, provided key—"dramatic," in the words of some scientists—support for the authors' idiotypic mimicry, or copycat, theory, because the foreign gene does not produce IgG. Those idiotype-positive antibodies had to have been made by the mouse's own genes, which did produce IgG.

In Table 3, David Weaver of Baltimore's molecular biology lab examined thirty-four of Imanishi-Kari's hybridomas from the engineered mouse and found that most of them did not show evidence of the transgene's RNA, an indication that any 17.2.25 idiotype in those cells had not been produced by the transgene.

Table 3, which tested relatively few selected hybridomas, was actually done before Table 2, which Baltimore described as a more global look at what was happening to the mouse's immune system.

Weaver's Figure 4 in the paper also was an important cross-check, or confirmation, of the paper's conclusions. The figure was said to be a photograph of an X ray from a test called a Northern blot. In this experiment, RNA is placed in the lanes of a gel, which is then charged so that the RNA of each cell's allotype travels down the lane. The gel is then bathed in a radioactive probe—a reagent made up of a piece of DNA—and an X ray is taken of the gel. The allotypes then can be distinguished by their size on the X ray. They look like smudges.

According to the paper, only four of the cells reacted to the probe for the transgene RNA, so Weaver then determined by the size of the RNA allotype that three were gamma, from endogenous genes, and one showed a band the size of the mu allotype of the transgene. The figure thus showed that most of the tested cells did not produce the transgene, and three of the cells that showed some transgene properties actually were endogenous.

Although the paper in its summary and concluding discussion clearly favors idiotypic mimicry, the discussion section does run down a number of other possible theories for what the scientists had seen. In fact, Baltimore says now that he never supported the copycat theory and the summary, limited editorially by space, is "not significant."

The *Cell* paper concluded by citing unpublished experiments by Moema Reis, Thereza Imanishi-Kari, and one "M. O'Toole" as further support for its own findings. Those 1985 experiments had indicated the presence of activated endogenous white cells in the engineered mice.

Imanishi-Kari would later say that it was O'Toole's unpublished experiments that tipped the *Cell* paper toward idiotypic networking. But "M. O'Toole" was surprised to see the reference. She had broken with her boss over those experiments, when she could not duplicate the Bet-1 tests and began to question the central conclusion of the *Cell* paper.

Batman and Robin

■

Walter Stewart sat in his attic lab at the National Institutes of Health in Bethesda, Maryland, staring at hundreds of bottles of exotic snails.

The experiment had to do with studying the structure of cells; Stewart was distracted from it on a spring day in 1986 by the ringing of the telephone. Stewart wasn't particularly surprised that he had never heard of the man who identified himself on the line as "Charles Maplethorpe of MIT."

It had been a busy time for Walter Stewart, and for his partner and nominal boss, Dr. Ned Feder. Although they were supposed to be focusing on their basic research on the genetic control of cell structure by studying the nervous systems of snails, Stewart and Feder were deeply embroiled in something much more enjoyable—a three-year controversy concerning scientific fraud at the prestigious medical schools of Harvard and Emory universities. The snail study was part of their work of basic research; cracking cases of scientific fraud was more of an avocation.

The Harvard-Emory fraud had really caught their attention. There was no question about whether cardiology researcher John Darsee had fabricated research: An NIH committee had concluded that he did. But what fascinated Stewart and Feder was that Darsee had numerous co-authors, many of whom were prominent scientists, on his long string of tainted publications.

So rather than concentrate on Darsee, Stewart and Feder used the voluminous record of the case to explore the performance of the forty-seven other scientists and researchers who at various times had collaborated with Darsee on published papers and abstracts. Had they been totally ignorant of the scam? Stewart wondered how it could be that men of science could be so easily fooled.

Of the forty-seven co-authors, Stewart and Feder concluded that thirty-five had committed sins of various gravity: failing to preserve original data; putting their names on papers for which they had done little work; and republishing old data in new forms without acknowledging that they had done so.

But Stewart and Feder also found that some of the scientists had published statements they should have known were false and had misdescribed how experiments were being done. In 1983, after reviewing eighteen papers, eighty-eight abstracts, and three book chapters, the two NIH scientists set out to publish their report in a scientific journal. Their reputation as troublemakers in the scientific community developed shortly after their preliminary draft was circulated among dozens of senior scientists. Many of the scientists thought the paper had strong merit, but then Stewart and Feder, the journals, and NIH itself were hit by a storm of lawyers' letters. In the wake of what had happened to Darsee, some of the scientists had retained lawyers. Their attorneys let it be known that any publication of the Stewart-Feder paper might well be greeted with a libel suit. One lawyer went so far as to warn that publication in any science journal of national circulation could constitute a separate cause of action for libel in each of the fifty states. Journals feared legal reprisal and rejected the report. Michelle Press of *American Scientist* magazine wrote Stewart and Feder: "We see no feasible way to use it. . . . Both our publisher and our long-time legal adviser believe that the legal costs incurred would be prohibitive."

But in their exhaustive manner—they knew no other way—Stewart and Feder would not let the paper drop. They asked for suggestions to improve the report, they made revisions, they wrote letters back, they haunted the scientific journals. They pestered and would not be pushed aside. Still met with rebuffs, Stewart contacted Floyd Abrams, the nation's foremost attorney on First Amendment issues, and asked his opinion on how damaging a lawsuit would be, with particular emphasis on how much defending even a spurious libel suit would cost.

Abrams, a partner in the prestigious New York law firm of

Cahill Gordon & Reindel, could not assure them that they would, in fact, not be sued. The state of libel law was in flux. Freedom of the press was taking a battering during the years of Reagan-appointed federal judges. The situation had become so bad, Abrams noted, that one small paper was forced to pay $9.2 million in damages as a result of investigative work on articles that were never even published.

Although he couldn't encourage Stewart and Feder to publish, Abrams did mention their experience in a piece published in *The New York Times Magazine* in 1985, to illustrate the need for change in the nation's libel laws to make it harder to intimidate writers. Abrams wrote:

"Mr. Stewart and Dr. Feder's manuscript has never been published in the two years since it was written. Several dozen distinguished senior scientists were given a preliminary draft. Most praised it. But an avalanche of lawyers' letters, some more than fifty pages long, descended upon Mr. Stewart and Dr. Feder, the N.I.H., and the small scientific journals to which their article had been submitted. Written on behalf of some of Dr. Darsee's co-authors, the letters threatened libel actions if the article was published. Each time Mr. Stewart and Dr. Feder revised the article and advised the lawyers, new letters arrived expressing dissatisfaction with the scope of the changes. . . .

"I learned of the existence of the unpublished manuscript when Mr. Stewart and Dr. Feder wrote to me asking if I could assure the editor of a small scientific journal 'that he will be protected against the possibly ruinous costs of a legal defense if he should be sued.' It was 'our impression,' they wrote, 'that this is what concerns journal editors, not the possibility (which we have taken great care to minimize) that our report contains false statements which could make a libel suit successful.' I could give them no such assurances."

The attention given the letter to Abrams brought Stewart and Feder an invitation to testify on libel law at a February 1986 hearing by a House panel on civil and constitutional rights.

"In the spring of 1983 we thought we had a way to measure how commonly professional misconduct had occurred in a large sample of scientists," Stewart told the panel.

"We found that more than half the scientists in our sample (thirty-five out of forty-seven) had engaged in questionable practices."

Stewart then told the committee of their discouraging and persistent attempt to get their conclusions published.

"We sent our report to fourteen other scientific journals (after being rejected by two prominent magazines) for informal consideration by the editor. We described to each editor the specific difficulties we had experienced. . . .

"Although more than half the editors commented favorably on our report," Stewart said, publication appeared unlikely.

It was only the power of *The New York Times* that eventually got the report released to the public. Through a Freedom of Information Act request, the NIH was forced to release the Stewart-Feder draft and hundreds of pages of related legal and scientific correspondence corroborating their conclusions. (The prominent British scientific journal *Nature* eventually published it January 15, 1987.)

That release triggered another *Times* article about their exploits, and the controversy was reported by the newspaper on April 22, 1986, under the headline "Major Study Points to Faulty Research at Two Universities."

The brouhaha over their paper led some wags to nickname the men Batman and Robin. But Dr. Ned Feder, a white-haired Marcus Welby lookalike, was an unlikely candidate for controversy. A quiet man in his late fifties, he calmly went about his work as chief of Biophysical Histology for the NIH's National Institute of Diabetes and Digestive and Kidney Diseases throughout the Darsee controversy. Feder earned his medical degree from Harvard and received a one-year research fellowship in 1954. Although he had planned a traditional career in medicine, he caught the research bug. In 1967 he left for the NIH, where he was joined a few years later by Walter Stewart, one of his former students.

As unassuming as Feder was, Stewart was a brash, enthusiastic science nerd who seemed to thrive on controversy. Even back in the seventh grade in his private school in Manhattan, Stewart was not willing to let the status quo rest. He reformed the show-of-hands system of voting in class elections by building a voting machine.

Although he was excited by math in high school and knew from his early days as an undergraduate at Harvard that he would be "in science," Stewart did not follow a traditional scientific career path. After being graduated summa cum laude in physics and chemistry from Harvard, he was appointed to the Harvard

Society of Fellows, a program under which one can follow a course of study without being required to seek an advanced degree. He also studied at Rockefeller University in New York. Stewart never did get a doctorate—a point his scientific critics would later note—but he published two important scientific articles after joining Feder at NIH.

In 1971, he had succeeded in isolating the structure of a particular tobacco-plant disease. These findings were published in *Nature*. Then, in that same year, he was asked by the journal's editor, John Maddox, to review a report Maddox had received from a scientist who claimed to have discovered that a chemical called scotophobin could retain learned behavior and be used to transfer it from one rat to another. Although other researchers appeared to have confirmed the bizarre phenomenon, it was still a highly controversial finding. Stewart's analysis suggested that the chemical did not exist.

His other major scientific achievement was the 1978 synthesis of Lucifer Yellow, a dye that can be injected into cells to outline their structures for study.

Later in the summer of 1987, while Stewart and Feder were in the midst of reviewing O'Toole's questions about the *Cell* paper, *Nature*'s Maddox asked Stewart for help on a manuscript on a discovery by an international team under Dr. Jacques Benveniste at the Inserm Laboratory at the University of Paris. According to the report, when water was exposed to a particular antibody, it retained a memory of that antibody and behaved as though the antibody were still present, even after being diluted to the point where there was no molecule left of the antibody. The article, signed by thirteen distinguished scientists from France, Canada, Israel, and Italy, would be an important buttress to the so-called science of homeopathy, which claims to treat disease with plain water bearing only the distant memory of some medicinal property.

Maddox would publish the article only after Stewart and the magician James Randi, who has made a career out of debunking psychic and religious fakery, studied it. After a week of experiments and study in France, Stewart and Randi concluded that the scientific team was guilty of self-delusion.

Stewart did not shy away from the controversial in his private life, either. He and his wife, Nancy, a Justice Department lawyer, caused an uproar by turning their six-acre lawn in Potomac, a ritzy

Washington suburb, into an uncut meadow. The overgrown property became a rustic nesting place for woodchucks, rabbits, deer, and foxes. Hummingbirds buzzed over the wildflowers. Their very proper suburban neighbors were horrified, to say the least.

Many neighbors apparently called Montgomery County officials to enforce the law against weeds. The Stewarts responded with a five-page "environmental impact statement" contending that their property may have "looked wild," but was actually safer than that of many of their neighbors, who were using chemicals to preserve their perfect lawns.

Initially county officials ordered the Stewarts to mow their yard. After Stewart threatened to test the case in court, the county backed down. The meadow stayed.

■

Despite the publicity over the meadow, it was the 1986 article in *The New York Times* that brought Stewart and Feder to Maplethorpe's attention. Feder and Stewart seemed like people who would be very interested in O'Toole's travails, Maplethorpe thought. He went to the Boston public library and thumbed through telephone books until he located a Dr. Ned Feder in Silver Spring, Maryland. Maplethorpe called that night but Feder's wife said he wasn't home, that he could be reached at the lab in the morning.

Maplethorpe quickly learned that one could start out meaning to talk to Feder, but would always end up hearing from Walter Stewart. Stewart, the junior member of the team, was far more loquacious than his partner, and Feder didn't seem to mind.

Maplethorpe outlined O'Toole's story in vague terms, telling Stewart that he thought the two scientists would like to keep track of a case study of an investigation into scientific misconduct as it was unfolding. But Maplethorpe wouldn't identify O'Toole and insisted that Stewart and Feder not interfere. O'Toole didn't know that he was contacting the NIH scientists. She would be angry that he had betrayed her confidence. "I will report back to you," Maplethorpe said.

"Why didn't you just complain?" Stewart asked Maplethorpe. This question surprised Maplethorpe. For two scientists who had such tremendous difficulty in getting their Darsee investigation published, Feder and Stewart seemed to show little understanding of why the authorities at MIT and Tufts were not sympathetic. Stewart and Feder seemed a touch naïve, but Maplethorpe wasn't disillusioned about going to them. Although they asked probing questions, they were still reassuring. "They don't act as if you're bizarre. They're very trusting and believing of the whistle-blower," he would later recall.

After O'Toole found the notorious seventeen pages of lab notes and got no satisfaction from Dr. Wortis or Dr. Eisen, Maplethorpe called Stewart back. Stewart was intrigued, but frustrated that Maplethorpe would not identify his colleague. Stewart was also insistent about seeing the crucial seventeen pages. "We have to have the data sheets," he told Maplethorpe.

"I can't get them; they're not mine," he replied. Although Maplethorpe's hushed voice on the telephone was a little disquieting, and Stewart was still too embroiled in the Darsee controversy to really focus on a new project, it wasn't in Stewart's personality to brush him off altogether. He encouraged Maplethorpe to keep him informed.

■

Then, at one of their lunches at a greasy spoon in Cambridge, during which O'Toole seemed terribly depressed about her confrontation with Baltimore and worried about her future, Maplethorpe took a deep breath and said, "There are some people who are interested in talking to you."

"Who? Those two guys at NIH?" Margot said sarcastically. She, too, had seen *The New York Times*.

Maplethorpe was silent; he wasn't brave enough to continue. At a later meeting later he steeled himself: "I should tell you something, Margot. I have talked to some other people—Feder and Stewart."

O'Toole was furious.

"Talk to them," he urged. It would be almost like therapy to talk to someone who trusts you, he said. That seemed to win her over. Talking to someone who would take her seriously and not treat her as an incompetent postdoc with an ax to grind was certainly appealing to Margot. She wasn't sure she would talk to the NIH scientists, but she finally agreed to let Maplethorpe give Stewart her name and the phone number of her mother.

Stewart called O'Toole in late June 1986. She sounded depressed to him also, as Maplethorpe had surmised. Although

Stewart felt he recognized in her voice the sound of a person who had tried to do the right thing but was overwhelmed by the establishment, he was a little worried that her supporting documents could be forged.

The only thing for him to do was to get a copy, and for him and Feder to look at the science themselves. But O'Toole still wasn't willing to turn over the seventeen pages. Stewart called again and again, lecturing O'Toole about her professional and ethical responsibility as a scientist. He even lobbied O'Toole's mother. What he didn't know was that it was Margot's mother who had actual possession of the material.

O'Toole had become obsessed by the seventeen pages and was constantly studying them to find some explanation of why she might be wrong or to find more evidence to prove that she was right. Her mother finally took the pages away from her; she lifted the burden from Margot and urged her daughter to relax and play a little. But Mrs. O'Toole wanted to send the documents to Stewart for his review.

Finally Margot did.

■

"I will always tell you the truth, but everyone else will tell you I'm lying," O'Toole warned Stewart.

Understanding the data was not easy, but Stewart and Feder were convinced that even journeymen scientists should be able, with study, to figure out the results of someone else's data—as long as those data were not forged. At the beginning, Stewart worried that O'Toole might have been a master forger. Her early statement to him that she would be accused of lying sounded a bit "loony" to Stewart, and kept him from immediately accepting the documents as real. But after many weeks of teaching themselves the science and poring over the data, Stewart and Feder became convinced that the seventeen pages were actual data for the published paper, as well as unpublished results. They came up with their own findings, and, as in their review of the Darsee case, Stewart and Feder soon found themselves in the middle of a scientific maelstrom.

They drafted their conclusions, contacted the original authors of the *Cell* paper for more information that might contradict their findings, circulated their paper among other experts, and tried to get their report published. It was the Darsee case all over again.

The Stewart-Feder report aimed for the heart of the *Cell* paper: the central finding that the results of the experiment appeared to "demonstrate unequivocally that the presence of the rearranged Ig gene in the germ line of the mouse in some way caused a marked expansion in the number of endogenous 17.2.25 idiotype-positive clones."

Although the talk around Cambridge and Boston was that O'Toole's documents might not be genuine data, Stewart and Feder showed that certain data were indeed included in the published results. While other data were not published, they said, "These data appear to have a direct bearing on the validity of the paper's conclusions." The *Cell* authors eventually acknowledged that the records did come from Imanishi-Kari's lab, although Baltimore would say later that O'Toole selected the particular records that would prove her points.

Using the measurements on a hybridoma from a transgenic spleen, the Stewart-Feder paper showed that according to the data actually published, the Bet-1 reagent was not particularly specific for the transgene. Although other scientists, under different testing conditions, have reported a greater specificity for the transgene, Stewart and Feder said the reagent as employed by the *Cell* authors did not do the job.

Stewart and Feder noted that lab records for a particular hybridoma, number 444 from the spleen, were published in Table 2, but actually confirmed that the reagent was not specific for the transgene. Table 2 showed the hybridoma expressing endogenous mu^b, but the seventeen pages indicated that the radioactive counts showing a reaction with Bet-1 (indicating the presence of the transgene) were actually higher than those for the reagent AF6, which was detecting endogenous mu^b. Nonetheless, the published table included number 444 as one of those producing the mouse's own mu^b. Stewart and Feder argued that arbitrary cutoff points for radioactive counts also affected the published paper's results for scoring particular cells as positive for the transgene idiotype.

They noted that, according to the lab records, the controls reported in Table 2 were not done at the same time as the other experiments. The actual contemporaneous controls were not reported, they said. The published paper also noted that only five of the thirty-four hybridomas listed in Table 3 were expressing the transgene, although unreported data in O'Toole's documents showed that twenty-two of the remaining twenty-nine that were purportedly not producing the transgene were actually reacting with the Bet-1, the reagent that detects the transgene.

"The authors' chosen methods of presenting their data convey an inaccurate and misleading picture of their experimental results," Stewart and Feder wrote. "The data the authors did not present were directly relevant to the accuracy of their assertions, and in fact, show many of the most important assertions to be false. Our analysis shows that their experimental results not only failed to support their main conclusions, but in many cases actually contradicted these conclusions."

The NIH scientists did not believe that their findings—based on the seventeen pages, the *Cell* authors' limited responses to the questions raised about their work, and one of the university reports—could be chalked up to differences of scientific interpretation. "We submit that it is simply a matter of scientific fact that the paper contains serious inaccuracies," they stated.

When Stewart and Feder initially contacted them by telephone and letter, Baltimore and his team—the co-authors designated him their spokesman—argued that the differences were indeed a matter of scientific interpretation. Stewart recalled that his conversations with Baltimore were anything but cordial, with Baltimore's voice rising louder and louder as they talked. Baltimore attacked O'Toole's motives and said that the Eisen and Wortis reports showed that her allegations were false. He also suggested that he would take legal action against Stewart and Feder, a threat Stewart wrote down in his notes of the conversation.

Stewart and Feder drafted their conclusions in the fall of 1986 and then went through the tortuous process of getting NIH approval to seek publication of their report. Written approval is required by NIH of all papers published by its "intramural" scientists. Although the approval is normally routine, Feder and Stewart couldn't seem to get anyone to sign off. The paper was bumped up to Joseph E. Rall, deputy director of intramural research at NIH.

Rall; a representative of the Office of Extramural Research and Training, which deals with NIH-funded scientists at other institutions; and other NIH officials met to discuss the discrepancies. The NIH officials were confounded by the Stewart-Feder position that they were obligated as scientists to publish their concerns, in order to caution other scientists who might be relying on the *Cell* paper, and by the possibility that the office might eventually need to conduct its own inquiry.

A Stewart-Feder briefing was scheduled for November 17. The two scientists were concerned, however, about how their report was being portrayed before the briefing. A November 5 memo from Mary Miers, who was "institution liaison officer" of the Office of Extramural Research and Training and was the official responsible for misconduct policy—referred to the upcoming briefing and two *Cell* articles that were "relevant to the alleged misconduct in science case recently brought to our attention by Dr. Feder and Mr. Stewart." In a response dated November 10, they stated: "We want to clarify a point to prevent any misunderstanding. As you know, we have made no allegations of misconduct. We notified your office of our interest in possible discrepancies between laboratories [*sic*] notes and the paper published in *Cell* because it seemed possible that there was misconduct and because we did not wish to compromise any actions you might wish to take in the future." It was a fine point, but one that was important to Stewart and Feder.

Rall informed the two scientists on November 13 that he had sent their manuscript to reviewers and asked for their continued cooperation. "I should like to stress particularly that there should be no communication with the co-authors until . . . extramural officials have determined whether an inquiry is warranted and the institutions have been so notified."

At the November 17 meeting, Stewart and Feder argued that it was appropriate for them to study the accuracy of the *Cell* paper, and that publication in a scientific journal would get the information out faster than an official review of possible misconduct. They came away from the meeting believing they had permission to continue their effort in their capacity as NIH scientists.

But they also had other problems to handle. Since the Darsee manuscript controversy, Feder had had to deal with unsatisfactory job-performance ratings. Official appeals were necessary to reverse them. Believing that the ratings constituted retaliation for their campaign to root out professional misconduct, Feder and Stewart wrote fellow scientists around the country for help. In the process, the two hinted at their study of the *Cell* paper:

"During the past few months we have carried out a study making use of the method of internal auditing suggested in the still-unpublished study [Darsee] mentioned above. Using this method we have analyzed an important paper in molecular biology recently published in *Cell*. The results of our analysis suggest strongly that assertions central to the conclusions of the paper are contradicted by the original data, and we have recently described these results in a form suitable for publication. Our findings are tentative; they need to be communicated to the authors of the paper for their response.

"We are writing to ask your help with our appeal of the unfavorable rating given to [Feder] by NIH. . . . Our supervisors sometimes maintain that our work on professional misconduct and its effect on the accuracy of scientific publications is trivial. It would be helpful to us, during the formal process of appeal, to have letters from outside scientists commenting favorably on the importance of our work on professional practices. If you believe this work is important, we would very much appreciate a letter saying so."

One of the scientists who responded was John Edsall, a professor emeritus in Harvard University's department of biochemistry and molecular biology. Edsall is one of science's heroes, a man who stood up to political interference with science in 1955 by declaring that he would accept no government research money that came with political ties. In early 1954, it was learned that the U.S. Public Health Service was denying or revoking grants for unclassified research on grounds that the scientists were disloyal or subversive. Edsall wrote at the time:

"Even if only a few of our colleagues are hurt—whether it is one person or many—I believe that we should stand up and protest on their behalf. In any case the threat is not to a few persons only; it is to all of us; for no one knows whether or not he will be the next victim, and whether or not he will find his own support cut away and his own future in jeopardy. . . . Under the circumstances I shall neither ask for nor accept funds from any government agency that denies support to others for unclassified research for reasons unconnected with scientific competence or personal integrity."

Continuing over the years to speak out on scientific ethics, Edsall was concerned about the possibility that work on scientific misconduct was hurting Feder's career. "I gather that the work on professional misconduct is not simply being rated as irrelevant in the evaluation of Ned's work as a scientist, but is actually being given a minus sign," Edsall wrote to Feder and Stewart in a November 26, 1986, letter. "Of course, I disagree fundamentally with such an approach, and would be more than ready to back you up in writing. . . . You have blown your whistle in relatively subdued tones, but many people are uncomfortable at being forced to listen."

The performance rating of "minimally satisfactory" was later changed to "fully successful" by Dr. Pierre Renault, deputy director of NIH's National Institute of Diabetes and Digestive and Kidney Diseases, who nonetheless stated that while he agreed with his supervisor's assessment of Feder's performance, he had decided it would be best to establish a new performance plan and evaluate Feder again later.

But getting performance ratings changed was much easier than getting approval for publication of the manuscript on the *Cell* paper. Rall's scientist-reviewers reported back with mixed feelings and urged that Stewart and Feder contact Baltimore and the other authors to seek more information and try to resolve the differences.

"I think the feeling is that your manuscript could well be an important study; however, it could equally well be that the seventeen pages you have are falsified or there may be additional unpublished data you are unaware of that substantiate the data in the paper and cast doubt on the data contained in the seventeen pages you have access to," Rall wrote in a December 12 memo to Stewart and Feder.

One of the reviewers noted that if O'Toole's documents were authentic, then "the validity of the conclusions in the resulting paper may have to be called into question. For example, Tables 1 and 2 in this manuscript are said to represent the published and the raw data versions of the same experiment, but are clearly discrepant."

Although, Rall said, Stewart and Feder should send a copy of the seventeen pages to the co-authors to ensure that they knew precisely what was being talked about, he added: "I am not sure that it is a wise move to ask to have access to all the authors' data so that you can perform an 'internal audit' on it." He informed them that he would be withholding approval of their manuscript for publication pending this attempt to work with the authors.

Stewart and Feder sent letters to the co-authors explaining their interest in the *Cell* paper and asking for access to the original data. Over the next several months there were letters and telephone calls back and forth, with many of the calls abruptly ended by one or another of the co-authors, Wortis, or Eisen. During one call, Stewart and Feder said, Imanishi-Kari noted that her work had already been reviewed by two committees, but added that there was a problem with the Bet-1 data. Despite Baltimore's public statements later, Stewart reported that in one phone conversation the scientist had referred to O'Toole as a "disgruntled postdoc" engaged in a personal vendetta. On January 21, 1987, they received a letter from Baltimore referring to a "discontented post-doctoral fellow" who had raised questions.

Baltimore sought to discourage Stewart's and Feder's inquiry, saying, "This is a dead issue," and would tell his colleagues so. "I do not recognize your right to set yourselves up as guardians of scientific purity," Baltimore wrote. He recommended that they focus on more "constructive scientific endeavors," a point he came back to in a 1992 interview in which he described their work as essentially "destructive."

During one phone call, Stewart claimed, Baltimore said that O'Toole had stolen the documents and called Stewart a vicious person. In a March 24, 1987, letter to Stewart and Feder, Baltimore wrote that their "attempts to quote back to me my own telephone conversation is [*sic*] in extremely poor taste and like your data, are selective."

Baltimore proposed that Rall appoint a panel of immunologists to determine if the *Cell* paper fell within scientific norms. But Stewart and Feder would have to promise that if the panel agreed with Baltimore they would never discuss the question again. Stewart suggested instead that Baltimore and his co-authors appoint a committee, that the two NIH scientists and the committee be given access to the original lab records, and that everyone be free to publish their findings. Baltimore also submitted his proposal to Rall. (In an April 2 letter to Baltimore, Rall characterized the proposal as a "very generous suggestion," which would be considered by Miers, the misconduct-policy officer.)

Two years later, in 1989, Baltimore continued to belittle the Stewart-Feder report as a faulty analysis based on just seventeen pages out of a thousand, written by two people without the expertise to understand the science. He was not surprised that the data in the seventeen pages appeared to contradict the published *Cell* paper, he told John Dingell's congressional subcommittee in hearings about the case. The seventeen pages represented data that O'Toole had selected to make her points, he charged, and were not representative of the entire study.

Baltimore also argued that the Stewart-Feder concept of an "internal audit" of another's scientific work was a disruptive interference in science. Indeed, he told Dingell's oversight committee that Stewart and Feder had conducted "so vitriolic" a campaign against him that he himself had called on the National Institutes of Health to investigate the *Cell* paper. He noted that he had received calls from fellow scientists as well as reporters about the unpublished report.

"On the basis of these seventeen pages, they wrote a long manuscript which attacked the *Cell* article," Baltimore testified. "They also began a campaign to discredit me and my work and to obtain evidence to support their faulty conclusions. This is a classic case of verdict first, evidence later. I refused to cooperate with them for two main reasons: First, they had already reached their conclusions; second, they were absolutely unqualified to evaluate the science of the paper."

Henry Wortis of Tufts was also quite dismissive of Stewart and Feder. In a March 2, 1987, letter to them, Wortis said "it would not be useful" to pursue their questions because they were not conducting a scientific study using scientific protocols. Wortis added, "There is no social or scientific gain in satisfying your curiosity." He noted that there were other laboratories studying systems similar to those written about in the *Cell* paper.

Herman Eisen, too, thought the Stewart-Feder audit was a waste of time, and in a May 11, 1987, letter to O'Toole indicated the depth of feeling the establishment scientists had against Batman and Robin. Oddly, his letter was a response to O'Toole's request for a copy of his memo summarizing his views of the dispute, which he had consigned to his own files and never sent to the university authorities. O'Toole wanted the memo because Eisen had told the two scientists that he would not release it without her permission, and because she had heard from Stewart that Eisen had referred to her as "pretty incoherent." Eisen explained in his letter that he was referring only to her initial verbal account of her concerns, when she appeared distraught.

Eisen said he would give her the memo only if she did not pass it on. Eisen wrote that he wanted to make sure the memo did not reach Stewart and Feder, "for I am, as a matter of principle, determined that in no way will I contribute to their 'pernicious and destructive' practice of trying, in the name of science, to engage many scientists' time and efforts in futile efforts to cleanse the scientific literature of all blemishes and imperfections." The effort spent on "Stewart's and Feder's obsession" wasted everyone's time, Eisen said.

In April 1987, after meeting with Rall and other top NIH officials, Stewart and Feder were prohibited from submitting their manuscript to any scientific journal. Miers, as the NIH misconduct-policy officer, would contact authorities at MIT about the questions. Rall sent the revised manuscript to four referees, three of whom later recommended against publication.

However, one reviewer, who eventually opposed publication, said that the manuscript made several valid points, including the matter of misrepresentation of the Bet-1 reagent's capabilities. "Therefore, the origin of the data in the bottom panel of Figure 1 showing complete absence of binding of Bet-1 to the b hybridoma product that is idiotype positive is dubious. A second point is that selection of hybridoma lines chosen as idiotype positive or negative is arbitrary. It is true that there are no data shown to support the cutoff values chosen," the anonymous reviewer said. Furthermore, the reviewer stated that if the Stewart-Feder paper accurately reflected the lab records and "if thirty-nine percent of hybridomas from a normal mouse are positive with this reagent, it obviously recognizes an isotype or a major subgroup specificity. If this accusation is true and data of this type were suppressed, this represents a serious ethical lapse."

But, not convinced "that this is true," the reviewer opposed publication. Obviously sharing the feelings of many scientists being drawn into this controversy, the reviewer wrote: "It must be asked what positive purpose would be served by exposition of these possible lapses in a manuscript from the laboratory of a famous scientist. . . . It is my opinion that the possibility is remote that anything more than some misguided enthusiasm on the part of junior staff and sloppy editorial procedures on the part of senior staff is involved here."

In the meantime, Stewart and Feder sent letters to other scientists, much as they had in November when they sought help in reversing Feder's poor job performance rating, asking for advice in dealing with their manuscript and NIH's prohibition against publication. They gave their colleagues a fairly detailed rundown on the *Cell* paper and O'Toole's travails, as well as describing their own efforts to talk to the paper's authors.

Word of the controversy was clearly getting around, despite Stewart and Feder's inability to publish formally. In their frustration, the two asked the American Civil Liberties Union for help, and the ACLU asked the tony law firm of Morrison & Foerster to represent Stewart and Feder. On July 14, attorneys G. Brian Busey and Edward T. Wahl wrote to the NIH's legal adviser, Robert Lanman, urging NIH approval for the manuscript. "We believe that NIH's continued refusal to allow Dr. Feder and Mr. Stewart to seek publication of their article constitutes an unlawful prior restraint," the lawyers wrote, adding that they believed that Feder's unfavorable progress review might be related to his efforts to seek permission to publish the manuscript.

Three days later, Stewart and Feder received a memo from Rall informing them that he had discussed the issue with James Wyngaarden, who was then the NIH director, and that it would be reasonable to allow publication. "The reason for the above decision," Rall wrote, "is the overriding importance of permitting free and untrammeled investigation and reporting thereof in any research institution."

But publication still was far from certain.

■

Stewart and Feder sent the manuscript on September 30, 1987, to Benjamin Lewin, editor of *Cell*. "We realize that you may not be overjoyed to receive this manuscript but we trust that you will give it your usual prompt attention," they wrote.

Lewin responded fairly promptly, but disappointingly. The format, he said, was inappropriate for the journal. He could not ask referees to review the data, because the data were incomplete. Further, *Cell* was not the right place for extensive analyses of fraud. The "sensible" solution would be for an impartial committee to review the report, something that Lewin noted that Baltimore said he proposed. If a correction should be warranted, *Cell* would publish one, Lewin added.

But Stewart did not let the matter drop. He wrote back noting that the manuscript had not mentioned fraud as an issue. But Lewin was not moved.

Stewart then sent the manuscript to Dr. Patricia Morgan, managing editor of *Science*, a respected journal published by the

American Association for the Advancement of Science. They complimented her on her reported remarks to the effect that editors should aggressively expose errors and fraud.

But Morgan, too, declined to publish. "The manuscript is not a scientific paper and cannot be peer reviewed," she wrote back. "The type of allegation you make is best handled by an investigating committee."

The indefatigable Stewart tried to persuade her as well as *Science*'s editor, Daniel Koshland, Jr., but to no avail. The British journal *Nature* also declined to publish.

Not until June 27, 1991, after the NIH Office of Scientific Integrity determined in a draft report that some of the *Cell* data had been fabricated, did *Nature* publish the Stewart-Feder manuscript. An accompanying statement said that the piece was being "published for completeness."

But in late 1987, even though NIH had begun to poke around into the case by asking MIT and Tufts for their reports, Stewart and Feder were becoming discouraged about their paper's prospects of formal recognition. Then the telephone in their cramped lab rang again with a surprising request.

Peter Stockton, a fierce investigator on the staff of an even fiercer congressman, John Dingell, wanted to talk to them about possible cases of scientific misconduct and the response of the scientific community. Would they come out of NIH's offices and meet with him and his colleague Bruce Chafin on Capitol Hill?

FOUR

Big John and the Subcommittee

■

In late 1987, Walter Stewart was probably the man most hated by the scientific establishment, but Peter Stockton was quickly gaining on him. Friends and supporters of David Baltimore and Thereza Imanishi-Kari would learn it was far more politic to vent their anger at Stockton, and Stewart, than at Stockton's boss, Representative John Dingell, the politician who would turn the *Cell* paper into front-page news across much of the country. One could hate John Dingell and think him terribly unfair—and many who suffered his searching investigations did—but few people had the fortitude and ego to take him on publicly.

Dingell, a Democrat from Michigan, was chairman of the House Energy and Commerce Committee, the oldest committee on Capitol Hill (although its energy jurisdiction is relatively recent) and—a fact lost on few in Washington—the former power base of the legendary House speaker Sam Rayburn. However, while John Dingell harbored an interest in becoming speaker, he always sounded perfectly sincere when he claimed to be just as happy as chairman of the Energy and Commerce Committee. And why not? Many would say that his committee chairmanship gave Dingell a far greater reach than the post of speaker ever could. The less cynical might attribute his attitude to the New Deal heritage he inherited from his father.

John D. Dingell, Sr., the son of a Polish immigrant who had changed the family name from Dzieglewicz, had a shaky start. He was born in poverty and suffered from ill health all of his life. His mother died when he was young, and Dingell Senior quit school at sixteen to start hustling newspapers for a living.

He worked a number of stints at *The Detroit Free Press*, where he was one of the founders of the typographical union. When he came down with tuberculosis, the doctors gave him no hope of surviving. With fresh air the only known treatment at the time, Dingell went to the Union Printers Home in Colorado, where he defied the doctors' predictions and lived. After leaving the home, Dingell worked for the telephone company out west, where he met and married Grace Bigler. John Junior, known to his family as Jack, was born on July 8, 1926, in Colorado Springs.

The elder Dingell always wanted to go back to school, but never was able to take the time. Earning a living during the Depression took every spare moment. He now supported his family by taking such tough jobs as putting down pipeline in the rugged Colorado wilderness. Jack's younger brother, James, born five years after Jack, said their father educated himself by reading the dictionary, page by page. "Dad was brilliant, with a wonderful common sense," Jim said. "He always wanted to get on and get a degree."

The Dingells moved back to Detroit, where, one day in 1932, the elder Dingell read a newspaper article declaring that "a new group of men would be sitting in Congress." Being a congressman seemed a lot better than the difficult jobs to which he had become accustomed. The more John Senior thought about the possibility, the more excited he became. He borrowed money from family members, ran for office, was elected in the sweep that carried Franklin D. Roosevelt into the presidency, and went on to become one of the great New Deal lawmakers, representing Detroit's West Side for twenty-three years.

Dingell worked his way up through the seniority system and became the dean of the Michigan congressional delegation. More important, he became the ranking Democrat, after the chairman, on the Ways and Means Committee, the House's ever-important tax-writing panel. The costs he had incurred in his fight with TB made him a strong believer in socialized medicine. He persistently—albeit unsuccessfully—pushed in Congress for compulsory national health insurance to be paid by payroll taxes. He was also an advocate for the construction of the St. Lawrence Seaway and a network of superhighways for the nation. While he was a supporter of things practical, Dingell fought for increased federal support of basic science research, and backed the expansion and consolidation of the various federal science research institutes into the National Institutes of Health.

Although the Dingell family was rooted in Detroit, Jack, Jim, and sister Jule grew up as children of the capital city. John Junior, particularly, soaked up the political nature of Washington, serving at one point as a congressional page. He went to school at the Jesuit-run Georgetown University, earning first an undergraduate degree in chemistry and then a law degree.

It was only when John Junior returned to Michigan as assistant prosecutor of Wayne County from 1953 to 1955, in which post he functioned as his father's "eyes and ears," that he began to absorb Michigan state politics. Representative Dingell needed his eldest son to touch base and keep him informed of what was going on in the 15th Congressional District, because his chronic ill health prevented him from traveling much to Michigan.

On September 20, 1955, John Senior checked in at Walter Reed Army Medical Center in Washington for routine tests related to his asthma. That night he died of heart failure at the age of sixty-one.

His vacant seat now had to be filled by a special primary and election. Dingell's friends put out the word that the veteran legislator had hoped that his twenty-nine-year-old son John would succeed him. Brother James briefly considered politics also, but recognized that John was the lawyer and had made himself well known to the people who counted in the home district. Jim, five years younger, also would have been too young to succeed his father that year. While Jim had also studied at Georgetown, he had gone the science route, earning his masters and doctorate in chemistry and going to work at the National Heart Institute as a research scientist. "I had fallen in love with biochemistry," Jim recalled—adding, however: "At the undergraduate level, [my brother] was a better chemist than me. He's no stranger to science."

There were still hoops for John Junior to jump through before he could succeed his father; numerous political hopefuls had been eyeing the seat for several years. But John trounced a dozen Democrats in the special primary and then went on to win the seat. He easily won his elections since, eventually becoming the sixth most senior member of the House.

Dingell's official home is now in a Detroit suburb, Trenton, but his congressional district is still one of the most industrialized in the nation, including part of Detroit and Dearborn, home of the Ford Motor Company, the Rouge auto plant, steel mills, foundries, and the nation's largest single community of Arab-Americans.

John Dingell has remained a man of interesting complexities. An enthusiastic hunter of elk and turkey and supporter of the National Rifle Association, he was also the author of legislation protecting endangered species and marine mammals. A true advocate of health care and hazardous-waste cleanup, he opposed Clean Air Act restrictions, which he says would hurt the auto industry. A fighter for consumer protections, he opposed airbag rules, again because of the auto industry's opposition. A quakingly intimidating bully in hearings, he was a divorced father who rushed home to take care of his four children, of whom he had custody. Dingell then wooed Deborah Insley, a lobbyist at the time for General Motors and an heiress to the Fisher Body family fortune. Insley was many years his junior; she finally agreed to a date after many rebuffs, when the congressman proposed an outing to the ballet at the Kennedy Center. Claiming that she saw a sensitive side unknown to most people on the other side of the witness table in the Energy and Commerce hearing room, Insley married Dingell.

The secret of Dingell's success in Congress, according to his former House committee colleague Timothy Wirth, a Colorado Democrat, was his ability to "live with ambiguity" and understand that the world isn't simple.

He also learned the rules. Dingell became an expert in the rules that run Congress, knowing that it wasn't enough to have right on his side—or even to have the votes in favor of the substance of legislation—in order to win. Outcomes often depend on the arcane rules that govern how Congress deals with legislation.

The authoritative *Politics in America* says of Dingell: "His mastery of House rules and procedural minutiae is renowned, as is his maxim about the power that knowledge gives him: 'If you let me write procedure and I let you write substance,' he told the Rules Committee in 1982, 'I'll screw you every time.' "

Nonetheless, staffers say Dingell doesn't just go for the jugular without a care. He questions them and himself about what they are doing, and often he asks, before going out to chair a hearing—or after coming back: "Are we being fair?"

Dingell used the House rules to work his way up in the congressional hierarchy. While only a member of the committee, Dingell led a revolt of the subcommittee chairmen against his predecessor, the aged West Virginia Democrat Harley O. Staggers, Sr., and stripped the "innocuous" committee chairman of much of his authority to set the panel's agenda and hire staff. Dingell soon became chairman of the full committee, as well as of its Oversight and Investigations Subcommittee, a post he held after leading the revolt in 1981.

Although as chairman Dingell restored substantial legislative power to the full committee, he now had to deal with the legacy of his rebellion, as well as of his own encouragement of junior lawmakers. The committee included some of the most aggressive, talented Democrats in Congress, and Dingell sometimes had to make alliances with his Republican colleagues to maintain control.

The committee under Dingell boasts one of the largest and most technically expert staffs on Capitol Hill. The panel's jurisdiction, controlled in an octopuslike fashion by Dingell, is immense. To have jurisdiction is to have significant power over what legislation will be considered by Congress. As *The Washington Post* described it in 1983, Energy and Commerce is the place where "crucial decisions are made affecting everyone who breathes, drinks, eats, smokes, watches TV and movies, listens to the radio, drives, plays the stock market, needs medical care, pays for insurance, enjoys sports, worries about hazardous waste and nuclear power plants, buys faulty products, rides the railroads, or gets buried."

Or as NYNEX lobbyist Thomas Tauke, a former Republican member of the House from Iowa, has said: "John Dingell feels about his committee much as Lyndon Johnson felt about his ranch. Johnson didn't want to own the whole world, he just wanted to own all the land surrounding his ranch. Dingell doesn't want his committee to have the whole world, just all the areas surrounding its jurisdiction."

In Washington, the phrase "powerful committee chairman" is a cliché among news reporters: All committee chairmen are powerful because they control the movement of legislation, they decide the agendas of their panels, they determine who gets to use the committee as a forum, and they decide on the hiring of staffers, who do most of the down-and-dirty work. Lawmakers don't cross their chairmen lightly, because their very access to the legislative system is at stake. And members particularly fear to cross John Dingell because, in the oft-repeated words of one, "he remembers."

New York's James Scheuer, once the second-ranking Democrat on the full committee and chairman of its consumer subcommittee, learned that lesson quite well. He and Dingell were barely on speaking terms for years, having fought over Scheuer's plan to require the use of air bags in cars. When Dingell took over the full committee in 1981, Scheuer discovered that his subcommittee had been abolished.

The respect, fear, and loathing that Dingell has engendered in his friends, critics, and targets are clearly reflected in some of the nicknames pasted on him: Big John (he's a burly six feet, three inches tall), Big Bad John, Emperor of Big Science, Torquemada (*The Wall Street Journal*), Thug (*The Boston Globe*), Dirty Dingell (environmentalists were upset with his opposition to clean-air reforms), and Iron Pants (sometimes hearings are kept in session for hours on end without a bathroom break).

But for all Dingell's toughness, many House colleagues like him as well as respect him. His committee protégés and staffers are intensely loyal, and not just because they are intimidated.

Dingell has used his committee and its oversight panel to right wrongs as he sees them in the government bureaucracy, the military establishment, and the corporate world. In the pursuit of good, Dingell can be sharp-tongued, nasty, arrogant, tough, vindictive, and mean. And very smart. Nobody enjoys being called for a command performance before his oversight panel, and "Dingell-grams," the penetrating requests for information that the subcommittee sends out, have left many a corporate executive and government official queasy.

Dingell's probes into the Environmental Protection Agency and its handling of the Superfund hazardous-waste cleanup program led to the resignation of EPA chief Anne Burford and perjury conviction of assistant administrator Rita Lavelle. Lobbyist Michael Deaver's perjury before the subcommittee earned him jail time. It was the Dingell gang that shone the light on the $640 Navy toilet seat, on shady dealings by defense contractors, on unsafe artificial heart valves, and on second-rate prescription drugs. An investigation into university billing practices for research grants led Stanford University to admit to charging the government for such "indirect" research costs as a wedding reception; Stanford president Donald Kennedy eventually resigned.

Dingell's investigations and public hearings almost always go after big names. He has never been one to be satisfied with merely nailing underlings. But there has been an embarrassing misstep or two. Dingell once led a probe into whether John Gibbons, a top officer in the prestigious investigative firm of Kroll Associates, had been misrepresenting himself as working for the subcommittee. At the time Kroll Associates, known for its investigative expertise in corporate takeover battles, was working for Drexel Burnham Lambert, and Dingell had been investigating Drexel and its role in junk-bond-financed takeovers.

While his colleague was under attack at a March 1989 subcommittee hearing, Bart Schwartz, Kroll's senior counsel, realized that his firm needed help. During a recess, Schwartz called Frank Mankiewicz, vice chairman of Hill and Knowlton Public Affairs Worldwide and an aide to the late Robert F. Kennedy. A Hill and Knowlton team began advising Kroll Associates on how to get its story out to counter the leaks from Dingell's staff. Pulling together a team of fifty to defend the company against the subcommittee, Kroll Associates combed through telephone records and reconstructed business schedules and then laid out its case to Dingell. From the D.C. law firm of Brand & Lowell, the company brought in former House of Representatives general counsel Stanley Brand and Abbe Lowell; the two advised on House procedures and what to expect in a hearing. Also brought in were William Geoghegan, a special counsel to the House ethics panel, and Gordon Hatheway, Jr.; both were partners in the D.C. office of the Pittsburgh law firm Reed Smith Shaw & McClay. Other lawyers were hired as well.

In the end, Dingell called off the probe into Kroll executive Gibbons. While he said there was some conflicting evidence, Dingell added, "One conclusion I am able to draw from your testimony is that you were acting in good faith." But what apparently motivated Dingell's retreat was possible evidence that his staff had engaged in questionable wiretapping to nail Gibbons.

Dingell said he believed the subcommittee had not violated any laws, but added that the staff's activities "did not meet the higher standards which this subcommittee has and must adhere to." In a private meeting, Dingell apologized to his committee colleagues; the aide who headed the inquiry soon left the staff.

But the Kroll case was a rare retreat for Dingell. Sometimes he didn't have to win to be successful or to maintain his power image. Representing the interests of the auto industry—his constituency—Dingell fought environmentalists for much of the 1980s over efforts to toughen the Clean Air Act, despite the measure's popularity. Then, in the 101st Congress, when it appeared a bill was inevitable, Dingell dropped his obstructionism and moved to get the best bargain possible for Michigan automakers. He fought, he objected, he compromised, he argued, he called in chits, but most of all he kept the bill moving through his committee. And in the end, the revised clean-air legislation was strong enough for environmentalists to support, while still more favorable to Michigan than originally expected.

■

Another source of Dingell's power in Congress is his staff, among the most expert and efficient on the Hill. The oversight subcommittee staff, known as "a kennel of junkyard dogs," is devoted to Dingell and his causes. They, too, are tough, skilled investigators who won't be intimidated by colonels, administration officials, or scientists. They have disdain for their former colleagues who cashed in to become lobbyists and members of the "Dingell bar," a group of lawyers who because of their experience on the staff or connections on the Hill are sought after by recipients of Dingell-grams. "We kick their butts," says one Dingell aide.

Following Capitol Hill custom, Dingell aides generally talk only off the record and are immensely helpful to newspaper and television reporters, with whom they share information. Lawmakers' periodic shock and outrage over leaked information to reporters are disingenuous—many, if not most, legislators and Hill staffers leak information to some degree. Well-placed leaks are part of Dingell's oversight subcommittee's process and power. When Dingell was chairman of the Energy and Commerce Committee's Energy Conservation and Power Subcommittee, that panel also freely used media leaks, an activity that came to light in testimony during a 1982 libel suit by Mobil Oil Corporation's president, William P. Tavoulareas, against *The Washington Post*. Testifying under subpoena, Dingell's investigator Peter Stockton told how he had provided several documents about the business dealings of Tavoulareas and his son, Peter, to *Post* reporter Patrick Tyler. The information formed the basis of Tyler's story detailing a letter sent by Dingell to the Securities and Exchange Commission about the Mobil president. Stockton had hoped to garner some publicity for potential hearings on the Tavoulareases' business relationships.

"Whenever we're getting near the hearings, we like to release some information," Stockton testified. "Whenever you release some information to the press, you're always after some publicity."

During a deposition in the suit, Stockton said that he was often a source for reporters in Washington. And some reporters gave him leads, often with the hope that Stockton's investigations would result in news that would get leaked back to them.

Clearly one of the oversight subcommittee's junkyard dogs, Stockton expressed enthusiasm for his job, which he saw as a combination of doing the right thing and having a good time. He thrived on controversy and was intensely loyal to his boss.

"It's the best act in town. It's fun," said Stockton. "If you have a view of the public interest and want to pursue it, he's the best. If there are things wrong, you've got to right them."

Stockton was irreverent, spoke profanely, dressed casually, and according to former subcommittee staff director Michael Barrett, had "absolutely, totally illegible handwriting." Dingell didn't use Stockton's rough language and was less irreverent, but Barrett said that boss and staffer shared a willingness "to question the establishment."

"He hasn't lost the fire in his belly," Barrett said of Stockton. "He's ready to disbelieve the obvious. He's very dogged and he reads the documents."

Stockton, who got his master's degree in economics at Ohio State, came to Washington in the early 1960s as an economist with the National Education Association. After a stint with Lyndon Johnson's budget bureaucracy, he moved on to the House Budget Committee to work for Representative William Moorhead of Pennsylvania. Stockton handled defense procurement and became involved in the famous whistle-blower case in which Pentagon cost expert A. Ernest Fitzgerald was fired on the personal orders of President Nixon for telling the truth about the $2 billion cost overrun on the C-5A cargo plane.

At Dingell's oversight subcommittee, Stockton and colleague Bruce Chafin, both research analysts, were deeply involved in investigations into allegations of improper billing practices by General Dynamics (including charging the government $155 to board an executive's dog). Chafin, with an MBA from the University of Florida, had come to the subcommittee a couple of years before on loan from the General Accounting Office to help in the General Dynamics inquiry for two weeks, and stayed. Subcommittee staffers, however, juggled a number of investigations at one time, and Dingell in late 1987 unleashed Stockton and Chafin to nose around the scientific community. Although the subcommittee had held hearings several years earlier about "cooked" scientific studies, Dingell had decided that the National Institutes of Health had been a sacred cow, with too little congressional oversight, for too long. The federal government, at that time, was funneling more than $5 billion annually in biomedical research grants through the NIH, and Dingell wanted to know how well the NIH and the nation's universities were doing at responding to allegations of scientific fraud and misconduct.

With the NIH in an uproar over Stewart and Feder's Darsee and *Cell* manuscripts, and the pair's letters to other scientists around the country about their job problems and the *Cell* research, it was not surprising that Stockton and Chafin would hear that the NIH whistle-blowers knew about some controversy brewing in Cambridge. That was when Stockton called Stewart and invited him and Feder to come over to Capitol Hill for a brainstorming session. He explained that the subcommittee needed a specific case to highlight problems in handling scientific misconduct; that was how the panel approached any area that it was investigating. "We normally take case studies. You can't study policy in generalities," Stockton explained.

Happy to have someone interested in their research into misconduct, Stewart and Feder met with Stockton and Chafin in January 1988 in the House Energy and Commerce's large hearing room in the Rayburn Building. Stewart lined up six glasses of water, drank them, and filled them up again. And as he expounded on various cases, he stalked up and down the length of the hearing room. "He looks like a doofball," Stockton worried. But Walter and Ned made sense to the two Dingell aides.

One of the cases involved a Dr. Robert Sprague, a professor of psychology at the University of Illinois, who in 1983 had written a letter to the National Institute of Mental Health detailing his evidence that an associate, Dr. Stephen Breuning, had committed scientific fraud in NIMH-supported research on drug therapy for mental retardation. Stewart told of how it took several years of

Sprague's efforts and congressional and press pressure before NIMH finally completed its investigation and upheld his allegations. During that time, however, and as a result of his whistle-blowing, Sprague's work came under close scrutiny; and he had to struggle with the institute over his own federal funding. His support was at first deferred, and then reinstated at a reduced level. Stewart wondered aloud how many scientists in Robert Sprague's position, knowing the conflicts and difficulties that were bound to follow the exposure of a colleague, would have made their knowledge public.

But the case seemed too old to Stockton; it wouldn't make the hit that they needed. It would be too easy for the scientific establishment to say that it had already dealt with the Breuning and Darsee cases. "Everyone will say that's old days, old shit," Stockton said. "We need something more current."

At that point Stewart and Feder brought up David Baltimore. The newer case immediately got the Dingell aides' attention. Baltimore, a Nobel laureate, was a big name; MIT and Tufts were big institutions; NIH was foundering in its response—and up against the big names of science was a lowly postdoc whistle-blower, Margot O'Toole.

Stockton and Chafin thought the Baltimore case was, well, interesting, and would serve their limited purposes. They didn't have a clue, however, that a firestorm would ensue, and that the subcommittee would be immersed in the case for years. As Stockton recalled, "We were going to whop the thing; kick 'em around a little and go back to Defense. One goddamn hearing and they'd do the right thing."

The aides quickly learned that getting scientists in Cambridge and Boston to whisper to them was going to be easier than getting them to testify before the subcommittee. At first eager to talk about the *Cell* paper and laboratory politics, other potential whistle-blowers soon evaporated. "We had a lot of support. They said, 'We're right behind you,' " Chafin recalled. "We forgot to ask how far behind."

In the end, of those not directly involved in the case, only Harvard's John Edsall, the scientist who had spoken out nearly forty years before against government interference in science, was willing to sit at the witness table near Margot O'Toole and to testify in general terms about scientific fraud and how to counteract it. Charles Maplethorpe, who had worked in Imanishi-

Kari's lab and who had been the one to inform Feder and Stewart about the controversy, testified only under subpoena. The subcommittee also blanketed Cambridge and Boston with subpoenas for test documents, for Wortis's and Eisen's reports on their inquiries, and for letters between various scientists.

The subcommittee scheduled a hearing for April 12, 1988, on the ability of the NIH and research institutions to handle allegations of fraud and misconduct. The panel lost the race to be first on the subject: The House Government Operations Subcommittee on Human Resources and Intergovernmental Relations, headed by New York Democrat Ted Weiss, had also been looking into scientific fraud and misconduct and had heard about the Baltimore case. Diana Zuckerman, an aide to Weiss, was talking to Stewart and Feder at the same time as Stockton and Chafin were. Weiss held his hearing with Stewart and Feder on April 11, but—believing that to do so would be premature and unfair—he did not mention David Baltimore by name.

But even before Dingell's subcommittee had held its first hearing, the pressure of its interest in the subject was being felt by NIH.

■

In January 1988, NIH established an investigating committee to look into the *Cell* paper. One of the committee members was James Darnell, in whose Rockefeller University lab Baltimore had once worked; the two had since co-authored a major molecular-biology text. Another of the panel's members was Frederick Alt, a former Baltimore postdoc who had written numerous papers with his mentor. Robert Charrow, the former deputy general counsel of the Health and Human Services Department, remembered telling NIH that it should not appoint the two scientists because of their close links to Baltimore. But NIH went ahead anyway.

Someone at Tufts who was not involved in the controversy and did not recognize the names happened to see a memo about the appointments and mentioned it to Margot O'Toole. O'Toole apparently passed on the information to Walter Stewart, who argued with Mary Miers and informed the subcommittee staff. "The close professional association between Dr. Alt and Dr. Baltimore, and the close professional and presumably business associations between Dr. Darnell and Dr. Baltimore, raise a substantial question as to whether Drs. Darnell and Alt have a conflict of interest," Stewart said in a February 17, 1988, memo to Miers.

Miers disagreed. In a March 11 memo to Stewart, she noted that in peer review "a decision in any given case must depend on balancing the frequency and recency of collaboration and the conscience of the individual asked to serve as a reviewer." She also blamed the difficulty in selecting the panel "on the publicity engendered" by Stewart's efforts to get his manuscript criticizing the *Cell* paper published.

In an interview with Dingell's staffers shortly before the planned hearing, Miers again defended the appointment of Darnell and Alt. But soon after Dingell took an interest in the subject, NIH had a change of heart and reconstituted the scientific panel.

(At a talk with science writers the following year, Baltimore said that when Darnell and Alt informed him of the appointments, he told them not to sit on the panel. "It would have been the last thing in the world I would have suggested to NIH," Baltimore said. But he did not make a point of his objections to NIH officials: "I couldn't see myself picking up the phone and telling them who to put on their panel. I just had to believe that they had their reasons.")

In true Dingell subcommittee fashion, word about the upcoming April 12 hearing got out to the scientific and general press; the involvement of Baltimore's name, even though he would not be invited to testify, ensured that the stories would get good play. To be sure, any news article about O'Toole's questions of the *Cell* paper would anger Baltimore and the Tufts and MIT scientists. But an April 10 *Boston Globe* article by medical writer Judy Foreman enraged Baltimore. Stockton, who was quoted and named, used the f-word: fraud. Besides appearing in the body of the article, Stockton's words were also repeated in a large pull quote, so they were impossible to miss.

According to the *Globe*, Stockton explained: " 'It's hard to tell if it's error or fraud. At certain times, it appears to be fraud and other times, misrepresentation.' . . . Although both investigations cleared the Boston researchers, 'Dingell will characterize' the inquiries 'as piss-poor,' Stockton said."

Baltimore denied that there was any fraud or misconduct. He acknowledged, Foreman wrote, "that in the lab notebooks 'there will be data that contradicts a given published version. Nobody in biology would doubt that. But if we let Stewart and Feder use that as the basis for a personal inquiry, all we'll do is inhibit the free flow of publication. And where there is a real suspicion of fraud, NIH has established a process of investigation and I completely support that process.' "

Foreman could not reach O'Toole for comment at the time. But she ended her piece with Stockton characterizing O'Toole as having been left shaken by the experience. " 'Thereza demoted her,' he said of Imanishi-Kari. 'She became the mouse cage cleaner. She has now left science, and with a vengeance.' "

Baltimore and the others jumped on the Stockton quote as evidence that the inquiry was biased against them. They already knew they were at a disadvantage because they were not being given a chance to speak at the hearing. At the time, Baltimore's longtime counsel Normand Smith III, a partner in Boston's Perkins, Smith, Arata & Howard (now Perkins, Smith & Cohen), said: "The hearings were rather a surprise. [We had only] two-days notice. We kind of heard on the grapevine that we were being tried in absentia."

Stockton was upset at being quoted by name, which could hurt—or at least anger—his boss. He believed he was talking "on background only," as usual, but perhaps Foreman did not know the rules by which unofficial Washington worked. He also contended that he actually had been misquoted: He wasn't speaking for himself, but describing how various others had viewed the case: O'Toole had thought it was error, while Maplethorpe thought there was fraud.

But despite the uproar over the *Globe* article and the reluctance of the scientific community to get involved in a public hearing, Stockton and Chafin were pleased with their main witness: Margot O'Toole. "You had such a perfect whistle-blower in O'Toole," says Chafin. "She reeked of integrity."

But O'Toole wasn't happy about going to Washington and testifying in public. And she would tell the subcommittee so.

The First Hearing

■

At 10:10 A.M. on Tuesday, April 12, 1988, John Dingell banged down his oaken gavel and called the oversight subcommittee to order in the Rayburn Building's Energy and Commerce Committee hearing room.

In his booming voice, Dingell laid out in a fairly comprehensive fashion why he was holding a hearing entitled "Scientific Fraud and Misconduct and the Federal Response." In his opening statement, Dingell tried to quell the growing suspicions within the scientific community that he was part of Capitol Hill's "anti-intellectual crowd," which was targeting research funds for the next budget-cutting round. Dingell could have noted, but did not, that his brother, James, was a researcher at the National Heart, Lung, and Blood Institute.

In his best congressional baritone, Dingell announced: "The Chair believes [that] biomedical research has made enormous contributions to society and that the National Institutes of Health has played a major role in this endeavor." He added, "This committee and this subcommittee have for a number of years not only sought to foster scientific research but also to vigorously support the National Institutes of Health in the face of attempted budget and scientific research cuts.

"The committee cannot afford to divert precious dollars into areas of meaningless or fraudulent work, which are counterproductive and which on occasion threaten the public health and well-being. Therefore, this is a painful but I believe necessary hearing today," Dingell declared.

He said the number of fraud and misconduct cases coming to NIH's attention had been growing over the past several years, averaging more than two a month, and that the committee believed "we are only seeing the tip of a very unfortunate, dangerous, and important iceberg." He added that some experts suspected that the growing competition for federal research money and the pressure to publish or perish to move up within a particular university were fueling the misconduct.

But before going too much further in his opening remarks, Dingell noted his and the panel's support of whistle-blowers. He was quite clear: John Dingell would find out if anyone from NIH or Tufts or MIT in any way pressured or retaliated against any of the people about to testify. "There is a widespread understanding within the scientific community of the plight of whistle-blowers. This committee has a special affection for whistle-blowers in connection with defense matters and in connection with other matters. That affection will be manifest in our hearings today," Dingell said. "There are obvious conflicts of interest here; and there are pressures to contain fraud and horror stories within a tightly closed group. Although a number of prominent people have privately told the subcommittee staff about what is going on within the scientific community, many requested that we not make them appear publicly, for fear of reprisal by their colleagues or termination of their grants."

Dingell went on to say he was shocked to discover that the NIH relied on institutions to investigate their own people. "Apparently here we have the possibility of the fox actively investigating the chicken coop," Dingell said. And "the NIH oversight capabilities are hopelessly inadequate—a one-person office with only a secretary and an administrative assistant to assist the head of that office. The office has little will or capability to conduct an independent investigation of allegations coming to their attention."

And then Dingell came to the Baltimore case—which, he noted, was pending before the NIH. Knowing that he would be at a severe disadvantage if he started a debate on the fine points of immunology, Dingell said that the subcommittee panel was not trying to judge the scientific merits of the case, but instead was attempting to examine how the system responded to allegations of possible scientific error or misconduct. His assurances of fairness sounded disingenuous to Baltimore's supporters, who feared that the planned testimony was already stacked against Baltimore and Imanishi-Kari.

Dingell's Democratic colleague on the subcommittee, Ron Wyden of Oregon, agreed with his chairman's remarks and added that scientific fraud was more serious than other ripoffs. Wyden called for a new independent program for quality assurance in research.

"Fraud in scientific research is like using substandard materials in the foundation of a skyscraper. The entire structure is compromised by the unsound foundation and enormous effort is wasted," Wyden said. "Scientific research is an incremental process and each set of experiments is a building block in a larger structure. We can't afford either the social or financial cost of having such structures compromised."

Dingell then called Feder, Stewart, and O'Toole forward and asked if they wished to be represented by counsel. None did, and he swore them in, a procedure that seemed part intimidation, part insurance—as well as an important aid in producing truthful statements, since a sworn witness later found to have lied could be prosecuted. Although the threat may have been subtle, it was not an idle one. Lobbyist and former Reagan confidant Michael Deaver's downfall came as a result of perjury before the Dingell panel.

Feder and Stewart spoke first, with Ned pointing out that they were presenting their own views as scientists and not as spokesmen for the NIH, the Public Health Service, or the Department of Health and Human Services. Feder laid out the importance of basic research and why it should be supported even though practical uses might not be discovered for years. "One example is basic research done years ago by Dr. Bruce Ames on the genetics of the synthesis of histidine in a bacterium. This research, widely respected at the time for its quality, appeared to have no practical value," Feder said. "Some ten years later, it formed the basis of a remarkable advance: a cheap and rapid test to identify carcinogens."

Although he acknowledged that no one had accurate figures on the incidence of scientific misconduct, Feder explained quietly and simply the damage that such misconduct does. "The scientist who cheats in his research violates the basic precept of science, which is to find the truth and make it known," Feder said. "A scientist who cheats injures other scientists by misleading them, and he injures the public by wasting their money and sometimes by falsely claiming discoveries that directly affect the public health and welfare. Similar damage can be done by the scientist who, through irresponsibility or carelessness, publishes erroneous research or fails to correct it. Finally there is the monetary cost of what might be called 'junk research' that is known at the time it is done to be of little or no value."

After giving a quick rundown on the Darsee case, Feder left it to Stewart to relate the Sprague-Breuning case and then describe how they became involved with Margot O'Toole. "We first heard about the dispute from Dr. Charles Maplethorpe, who is here today. Acting entirely on our own initiative, we contacted Dr. O'Toole and requested a copy of the original laboratory records," Stewart said. "We emphasized our view that a scientist with private knowledge of serious defects in a published paper has a professional obligation to make those defects known."

Stewart talked about the paper he and Feder drafted on the basis of the seventeen pages of data and the difficulties they were having in getting it published. They had only recently received the Eisen and Wortis reports, Stewart added, and were analyzing them. "Our preliminary findings are that the reports contain facts which confirm some of Dr. O'Toole's assertions. The remainder of Dr. O'Toole's assertions are either ignored or treated evasively," Stewart testified. "We consider both reports so seriously defective that they are not useful analyses of the paper published in *Cell*. This is particularly disturbing in light of the scientific credentials and the distinction of the two scientists who prepared the reports."

Stewart noted that the scientists being criticized at the hearing were not there to defend themselves, and that a final evaluation of the events surrounding the *Cell* paper could not be made at that time. However, he listed "three disturbing issues that must eventually be considered besides the accuracy of the *Cell* paper itself." The three were the apparent inadequacy of the Eisen and Wortis reports; Baltimore's response to the allegations of scientific error; and the reluctance of many scientists to express their views on the accuracy of the Feder-Stewart paper, as well as the difficulties Feder and Stewart faced in publishing it. The two scientists' official testimony, their unpublished draft, and a detailed chronology of their dealings with Baltimore, Imanishi-Kari, Eisen, and Wortis were turned over to the subcommittee, which in a fine irony published them in the record of its formal proceedings.

It was only then that the subcommittee got to its star witness, Margot O'Toole, the whistle-blower who, the staff was convinced, "reeked of integrity." Told by the staff that she would be subpoenaed if she refused to testify, O'Toole reluctantly agreed to appear "voluntarily." Grim and more than a little frightened by the public spectacle, O'Toole started by telling the panel that she was not happy to be at the witness table. "Mr. Chairman and members of the subcommittee, my name is Margot O'Toole and I am appearing here today at your direction. I would not wish to volunteer this testimony as it involves many people I have admired all my professional life and some who were my friends.

"I have broken with these people over what I perceive as a major disagreement on matters of ethically acceptable professional practices and our responsibilities as publicly supported scientists," she went on, her voice quavering. "My dispute has halted my career, disrupted my social milieu, and had a devastating effect on my life."

Possibly still unable to face the totality of the break with her former colleagues, O'Toole added: "I know that it has also been extremely upsetting for those who disagreed with me." But, she continued, "we scientists should have been able to resolve our differences by ourselves in the academic tradition. I deeply regret that the matter has come to such an impasse. I would like to be able to forget my ordeal entirely and get on with other things."

O'Toole gained strength as she testified about the events that led to her break with Imanishi-Kari. She told how the results of her own experiments conflicted with Imanishi-Kari's and how, believing that she was in error, she continued to repeat them, taking up time and expensive laboratory supplies. Imanishi-Kari became impatient.

"She insisted that the discrepancies were the result of my incompetence and that I should accept as valid the findings as she presented them. This would have meant ignoring my controls and was therefore unacceptable to me for scientific reasons," O'Toole said. "Communication between Dr. Imanishi-Kari and me deteriorated steadily."

O'Toole noted that in the fall of 1985 she saw the manuscript that would later be published in *Cell* the following spring. She testified that she tried to discuss the discrepancies between her experiments and Imanishi-Kari's, but that her boss dismissed her results and maintained that she, Imanishi-Kari, had gotten the opposite results by doing the same work as O'Toole. But Imanishi-Kari would never show her the records for the experiments.

"As my knowledge of the available procedures in the laboratory increased, I began to doubt that certain other experiments could have been performed as claimed in the [*Cell*] paper," O'Toole testified.

Then—despite their bad relations and O'Toole's banishment to mouse breeding—Imanishi-Kari confirmed to her that some of the experiments had not been done as presented in the *Cell* paper, O'Toole testified.

Relating the story of finding the seventeen pages of data, she said the records "caused me to doubt that the underlying data supported the main conclusions of the now published study. In fact, it appeared that a number of the conclusions were actually contradicted by the records." She added that at the Wortis meeting with Imanishi-Kari, "my assertions that the data did not support the published claims were completely confirmed."

O'Toole testified that she told the MIT ombudswoman, Mary Rowe, that there were "serious errors" in the published paper and that she discussed the problems with Eisen, Baltimore, Imanishi-Kari, and Weaver. And again, she made clear that while she had aggressively worked within the scientific community to resolve what she thought were serious errors, she had not sought to take the dispute public.

"At this point, I had brought the matter first to Dr. Imanishi-Kari's friends and supporters, and then to her collaborators and officials at MIT. I believed that I had explored all my options and fulfilled my professional responsibilities. I would like to take this opportunity to state under oath that I did not know then, as I do now, that Mr. Stewart and Dr. Feder were being advised of these events as they occurred," O'Toole testified. "I can take neither the credit nor the blame that this matter has come to light. I dropped the matter completely and left science, saddened and disillusioned."

While she testified about serious errors and about experiments that had not been done as published, and while she harshly indicted the scientific establishment for being too forgiving of error, O'Toole never uttered the word "fraud."

Under questioning by Dingell, O'Toole allowed that MIT's Herman Eisen, who conducted the second review of the paper, implied that she was actually charging fraud. "I said no, I was just looking at the data, and he said that it looked like the same thing to him, and then after I gave him the memo, he called me and he asked me how could one explain these discrepancies without invoking fraud."

"And what did you say to that?" Dingell asked.

"I said that the judgment of the conduct issues was not my job," she replied. "I was a subordinate. My responsibilities were to the scientific accuracy, and I was very adamant about that, but I was only talking about the scientific content of the paper."

After leading O'Toole through a discussion of her meeting with Baltimore, Dingell called Charles Maplethorpe to come forth and be sworn in. Nervous about being the focus of attention, Maplethorpe felt disoriented. From the witness table, people look up to the dais; Maplethorpe, not quite sure where the voices were coming from, became confused. When John Dingell's authoritative voice boomed at him, "Do you desire to be advised by counsel during your appearance here?," Maplethorpe was thrown off.

Standing, waiting to be sworn in, Maplethorpe stammered, "Do I have the option—I'd like to have [the] option . . ." But Dingell couldn't hear him and told Maplethorpe to sit down and use the microphone. Maplethorpe looked over at Peter Stockton, whose face bore an expression of fear. Stockton was worried that his anxious key witness might "bolt." Realizing that he was acting inappropriately, Maplethorpe regrouped and said, "At this time, I waive the option of counsel."

"I don't think you have to be overly apprehensive about the committee's attitude towards you," Dingell said, trying to reassure Maplethorpe. The chairman then noted for the record that this witness was testifying under subpoena.

Representative Wyden led the questioning, asking Maplethorpe what led him to be suspicious of Imanishi-Kari. "Well, to begin with, I would say that my suspicions were originally based on her behavior in the sense that she was conducting the research in what I considered a secret[ive] manner, and would not share the data with other people in the laboratory, including myself," Maplethorpe testified.

While O'Toole had been cautious, staying away from charges of fraud, Maplethorpe and Wyden quickly got down to business.

"Doctor, did you observe what you felt was falsified and misrepresented data?" Wyden asked.

"Yes," Maplethorpe replied clearly, if briefly.

Under questioning from Wyden, Maplethorpe told of a June 1985 meeting between Imanishi-Kari and Baltimore's postdoc David Weaver, in which Weaver expressed concerns about the experiments and Imanishi-Kari told of problems with the Bet-1 reagent. That was the same reagent that O'Toole had problems with later.

"What was your reaction when the *Cell* article appeared with these various concerns unresolved?" Wyden asked. "You were shocked, were you not?"

"Well, when a train wreck happens, one is shocked," Maplethorpe said, "but if one sees the train wreck about to happen, I mean . . ."

Maplethorpe testified that he reported to the assistant to the MIT president that he suspected that Imanishi-Kari was committing fraud, but that he was simply handed a form detailing the institution's fraud guidelines. But he didn't take the matter any further; he simply wanted to get his doctorate and get out. "I felt there was no question, but that if I were to make a formal charge of fraud, that it would not be taken seriously and then I would be the person who would be worse off for it." Instead, after O'Toole's problems, Maplethorpe alerted Stewart and Feder.

Representative Gerry Sikorski, a Democrat from Minnesota, returned to O'Toole and asked about the consequences of raising her challenge. She said it was hard to think of some career other than science, and yet it was impossible to return to her original dream.

"To go back to science," she said, "I would have to go back as a person who had raised ridiculous and trivial and unsubstantiated allegations against my supervisor for vindictive reasons and this is how I was portrayed and this is how I was treated and I knew that if I ran a lab myself, I wouldn't want such a person in that lab." She testified that she tried to clear her name with Eisen and Tufts but got no satisfactory response.

The subcommittee moved on to John Edsall, who discussed the seriousness of fraud in science, and to Robert Sprague, who described the difficulties he ran into when he made allegations of scientific fraud against his co-researcher Stephen E. Breuning at the University of Pittsburgh. (Within a month of the Dingell hearing, Breuning was indicted for fraud; he later pleaded guilty to a felony.)

Edsall noted that many scientists believe that scientific misconduct happens rarely and that "fraud will be found out sooner or later, if the work is of any importance." He also suggested that some problems are due to competitive pressures to publish, and to the failure of senior scientists to take responsibility for the development of junior scientists.

And while noting that whistle-blowing is a professional responsibility, Edsall had discouraging words for the scientist who speaks out. He seemed to be speaking of Margot O'Toole's situation: "I think it's very hard for me to say how the plight of the whistle-blower can be made better except I think we've got to have increased respect for the people who seek to detect—who recognize fraud and misconduct in any area and speak up about it," Edsall said. "The sense of loyalty among members of a community tends to be very strong, and a whistle-blower is usually regarded as disloyal. That almost inevitably involves a painful conflict.

"I think that it's only fair to warn a potential whistle-blower that he or she is likely to face a great deal of trouble and distress. On the other hand, for some people, fortunately or unfortunately, the distress of not having obeyed one's conscience may be more troubling than the other difficulties they have to face later on."

The hearing, like most of Dingell's subcommittee hearings, wore on for six hours without a lunch break. The panel finally called on a group of officials from the Department of Health and Human Services and the NIH. The government team included Robert Windom, assistant secretary for health at HHS; William Raub, the deputy director of NIH; Mary Miers; Jesse Roth, the scientific director of the National Institute of Diabetes and Digestive and Kidney Diseases; and Rex Cowdry, the acting deputy director of the National Institute of Mental Health.

First, Windom was given a chance to make his formal statement about the prevalence of misconduct in science and how the government was handling such cases. While insisting that the vast majority of scientists believe in the rigorous conduct of experiments, Windom said "those who engage in inappropriate practices represent a significant threat to the integrity of science and the confidence of its public and private sponsors."

He noted that in 1981, after some incidents were widely publicized, NIH officials and other scientists had contended that the frequency of such misconduct was "vanishingly small." However, since then, Windom said, the NIH had handled fifteen to twenty allegations annually in its extramural programs, which support the work of about fifty thousand scientists.

In 1982 the NIH established an alert system, a procedure for the confidential sharing by grant-making officials of information about ongoing investigations. Then the NIH began to coordinate the development of policies and procedures for the entire Public Health Service. While the NIH continued to investigate allegations during this time, Windom noted that procedures for dealing with misconduct in grant and contract programs were approved as interim Public Health Service policy only in June 1986. At the same time, the Health Research Extension Act of 1985 required that any organization applying for research funding must provide assurances that it has established a procedure for responding to reports of fraud, and it must report the initiation of any such investigations to the secretary of health and human services. As Windom testified, a "Notice of Proposed Rulemaking," required before federal regulations interpreting the law could be implemented, was being prepared.

Windom contended that the PHS policies reflected the department's belief, and that of Congress, that universities and research institutions should bear the primary responsibility for preventing and detecting scientific misconduct in their programs.

"The attention given to a few celebrated instances in which institutions fail to conduct an adequate investigation, or at least provide an erroneous conclusion, should not overshadow the steady progress in both their ability and determination to carry out this important task," Windom told the panel.

PHS policy, Windom said, was that once the service received an allegation, the staff would check grant and contract records to ensure that the matter involved PHS funds; would consult with program staff on whether the matter warranted PHS attention; and would contact the inspector general if criminal activity was a possibility. If the PHS determined that an investigation was necessary, the institution involved was asked to conduct the inquiry and report back. After that inquiry was complete, the PHS determined if the case warranted entering the name of the accused into the alert system, or taking some other action such as delaying new funding to the accused. A review of the case was then undertaken by senior staff, who could recommend sanctions to the agency head.

Members of the subcommittee immediately jumped on Windom when he concluded his opening statement. It seemed clear that they had already accepted the testimony of O'Toole, Maplethorpe, Stewart, and Feder. As Maplethorpe later reflected—to his own surprise—the subcommittee's hearing was well scripted and little happened spontaneously. Or perhaps, like good prosecutors and defense attorneys, the panel members didn't ask questions they didn't know the answers to.

Oregon's Wyden was first. "Can you name three cases where universities or institutions thoroughly and accurately investigated and found serious misconduct or fraud within the university or institution?"

Windom seemed flustered. "Mr. Wyden, I'd like to refer that question to—I cannot cite you three cases. I do not have those cases in front of me."

Wyden bore down, however. "Could you cite one case where a university or an institution really got to the bottom of serious fraud and misconduct when it began?"

"Well, let me just reply, no, I cannot at this point give you one," Windom responded. "I would like to say, though, however, that of the one hundred and two cases that have been brought to our attention through the NIH investigative process, over sixty percent of those have been handled within one year's time and many institutions involved were able to handle those."

"But Mr. Windom," Wyden continued, "that's not my question. You have given a ringing endorsement of the current system, a system, that I must say, after what I've heard this morning, I find appalling because of its conflicts and the cronyism, and the forces that push people to look the other way. . . . Now you said that you couldn't give us even one example. How do you then make the assertion that the current system works so well?"

Windom started to reply: "Well, I think that the system is working well by the fact that there have not been cases involved in the sense of the progress, but I do not have . . ." But Dingell interrupted, in the highly stylized fashion of Capitol Hill: "Would the gentleman yield? Doctor, how can you say that the system is working well? All you can infer logically from the fact that you can't cite a case is that you don't know of any cases. You can't tell us that the system is working well."

Raub tried to help Windom out, but Wyden spoke again, battering Windom with the Darsee case, the Breuning case, and others in which universities had failed to conduct adequate investigations. Raub tried again, and again was ignored by Wyden. Finally, the lawmaker turned to Raub, who gave some vague instances involving Harvard, the Dana Farber Cancer Institute, and the University of Cincinnati—and then, to Wyden's exasperation, turned to Mary Miers for the details.

Wyden asked Miers for the number of cases in which the universities or institutions rooted out the fraud compared to the cases where they did not do so. She did not have the figures with her, so she reported to the panel later: "We have interpreted the words 'root out' to mean 'a discovery that resulted in an allegation of scientific misconduct.' Based on that interpretation, the following table portrays the sources of allegations from 1982–1987: Institutions: number 35; NIH/FDA, 17; Individual, 44; Anonymous, 8; Total, 104."

Representative Thomas Bliley, Jr., a Virginian who was the ranking Republican on the subcommittee, asked Windom about allegations that the research-funding system had been turned upside down and that an increasing number of scientists were conducting research solely to get NIH grants, rather than NIH awarding money for scientific research.

Windom handed off to Raub, who contended that there were so many ideas worth pursuing in science that he doubted second-rate projects, thought up simply to get money, would compete well.

"It's my belief, sir," Raub continued, "that the issue has been vastly overstated in terms of the behavior of the scientific community. There have been many instances where a matter of difference of technical opinion over the meaning of some data has been construed by some as fraud or deliberate misrepresentation. There have been other instances where individuals, in the judgment of others, have not always pursued the most rigorous measurement or may not have applied the correct statistical procedure. We have been careful not to label those as fraud, in that we see those more in the realm of the permissible differences of opinion among experts in a rapidly evolving subject area. None of us would presume that all scientists are saints, and therefore there inevitably will be instances of individuals who deliberately will mislead their colleagues, or indeed may be sufficiently compromised in their emotion or mental health to be able to distinguish proper conduct from improper conduct. It remains my firm belief that in those extreme instances we are dealing with a vanishingly small number, and that this is not a signal for any deep disease within the scientific community itself."

Raub's ringing defense was like a red flag to Dingell, who pressed him about the Darsee case. Raub acknowledged that the process had not worked well at Harvard, although he noted that in a subsequent case, the university acquitted itself quite well. The government recovered $120,000 in research funds in connection with the Darsee case and conducted an on-site review of the laboratory procedures of Darsee's superior. However, under questioning from Dingell, Raub admitted that Darsee's supervisor, Dr. Eugene Braunwald, received no sanction and continued to serve on an NIH advisory board.

The subcommittee eventually got back to the Baltimore case, which the NIH officials said was still open. "Oh, you regard it as still open? And how long has this review been open? It is now well over a year, is it not?" Dingell asked.

Mary Miers contended that the matter had first been brought to the agency's attention by Stewart and Feder as an issue of "possible scientific error and not as scientific misconduct." The decision to undertake a formal review, she said, was made only after reviewing the two scientists' manuscript analyzing the seventeen pages, as well as some institutional reports. Dingell asked for documents supporting Miers's contention and received the November 10, 1986, memo in which Stewart and Feder responded to Miers's memo describing two *Cell* articles as "relevant to the alleged misconduct in science case recently brought to our attention by Dr. Feder and Mr. Stewart."

In that memo, quoted earlier, Stewart and Feder explicitly said that they had not alleged misconduct; rather, they said, they had noted discrepancies in the data for which misconduct was a possible explanation. They added that they "did not wish to compromise any actions you might wish to take in the future."

It seemed clear to the subcommittee investigators, once again, that even though Stewart and Feder were being careful in their language, as O'Toole had been, they had said enough to alert the authorities to the potential seriousness of the problem.

For the government scientists, the hearing became an endurance test. Dingell pointed out that Imanishi-Kari continued to serve on an NIH board that reviewed the scientific merit of research applications. Miers tried to head him off by noting that the Imanishi-Kari case was an open investigation. That did not deter Dingell, who sarcastically asked, "Now is your test for getting on a board incompetence or misconduct?"

He then began to badger the witnesses over their standards for members of review committees. Miers said that NIH did not know about the MIT and Tufts reviews when they took place, so it could not have taken any action about who could conduct those reviews.

"What do you think of the investigations made at those two institutions at this particular time?" Dingell then asked. "Does it appear to have been thorough, fair, impartial, or does it appear to have other characteristics that are less desirable?"

Miers wanted to ask for advice from NIH counsel, but Dingell would not allow it. "Your counsel is here to protect your constitutional rights. The counsel is not here to advise the committee," he said. "If we have questions of counsel, we'll ask him. Now proceed to answer the question."

"Sir," Miers said, "I have heard accounts this morning that I find troubling."

At Dingell's request, she submitted the Eisen and Wortis reports—documents that O'Toole had had difficulty obtaining on her own—for the record of the subcommittee. But the subcommittee still was not letting Miers and the others go. Wyden and Dingell wanted to know how NIH had gone about establishing its investigating committee of three members, two of whom, Darnell and Alt, were close associates of Baltimore. Prior to the hearing—and since Miers's meeting with the subcommittee staff—Darnell and Alt had been removed by NIH, and a new panel was being formed.

Dingell told of one of the first cases he tried when he was "very young and very dumb. . . . The jury and I were going through the voir dire examination, which is where you try and find out whether the person is really qualified to sit on the jury. I said, 'Now is anybody here related to the defendant?' And two of his cousins stood up. I always thought, you know, if I'd let that case go to trial without asking that question, I wouldn't have gotten a very fair trial. I find it curious that on the panel that NIH and the HHS selected to review the behavior of Drs. Imanishi-Kari and Baltimore, you have three individuals, one of whom is a co-author of a textbook with Baltimore and another who has co-authored 14 papers with Baltimore. Now, first, would you say that those are people who have no personal or professional association with Dr. Baltimore?"

Miers, the misconduct-policy officer, took responsibility. "I didn't do my homework, sir." She conceded, however, that she had told the subcommittee staff a week earlier that she thought Darnell and Alt would have been objective, and, indeed, she still felt that way.

"Based on all of our information, these are objective, fair people," Miers said, noting that "there are enough people looking at this investigation that the probability of their being able to get away with anything is rather remote."

Under further pointed and sharp questioning from Dingell, Miers said she did not request or get the Tufts and MIT reports on O'Toole's allegations until after she had read the Stewart-Feder manuscript. After reviewing the reports, she then asked scientific staffers to review them; then, determining that the reports did not answer the questions raised by Stewart and Feder, NIH decided to convene its own investigatory panel. But, Miers said, she did not give the universities' reports to Stewart and Feder because they were not investigating the case as representatives of NIH.

Dingell was indignant. "This lasted over a year. You had the papers from Tufts and MIT for seven months. What did you do during either that year period or that seven-months period?" he asked. Miers tried to explain that NIH had decided to convene a panel, but Dingell would have none of it. "The fact of the matter," he said, "is you didn't do very much, did you?" He repeatedly asked Miers why she had never talked to O'Toole about the scientist's concerns, and repeatedly was dissatisfied with Miers's response that the investigation had not yet started.

Finally, after six intense hours, Dingell brought the subcommittee hearing to a close with a warning to the NIH and HHS officials: "Now I don't want you to get the impression that this committee is going to stage a one-day hearing, bring you up and make you miserable, and then let you go on about your business. When we do these things, we try to see to it that the pain and suffering goes on for a greater period of time, until the abuse that is obvious is addressed."

Indeed, Dingell was not going to let the subject drop. He got NIH to temporarily detail Stewart and Feder to the subcommittee to work on the panel's probe, and the next month he sent his investigators to Cambridge to at least try to interview the *Cell* paper's authors. More subpoenas were sent out to secure more documents, including all of Imanishi-Kari's lab records.

It was certainly clear to David Baltimore that he would have to take aggressive action to counterattack the Dingell subcommittee.

SIX

The Nobel Laureate

■

May is a beautiful time of year in Cambridge; the days are clear and warm enough to shake off the dreariness of the New England winter. Work in the scientific labs at MIT as well as the other research institutions, universities, and biotechnology companies in the Cambridge-Boston area competes with the spring afternoons in Kendall Square, Harvard Yard, and Boston's Faneuil Hall. But this day in May 1988 was not so beautiful for David Baltimore. Baltimore was perplexed and angry, even more so than he had felt a month earlier when he learned from a newspaper reporter that U.S. Representative John Dingell was going to hold a hearing—in two days—on allegations that a paper Baltimore had co-authored might have involved fraud.

But a newspaper article that appeared after the telephone call was even more of an affront. A *Boston Globe* article about the upcoming inquiry included his photograph next to a large-type quote from Peter Stockton: "It's hard to tell if it's error or fraud. At certain times, it appears to be fraud and other times, misrepresentation." Under Baltimore's photograph was the caption "His paper is challenged."

Sitting in his office at the modern Whitehead Institute for Biomedical Research, David Baltimore was enraged. How could anyone question his integrity, his motives? How—at a time when he was building up the Whitehead Institute after earlier accomplishments that could have led a lesser man to coast on his reputation—could he be dragged into a national scandal? He was on the short list of candidates to succeed Paul Gray as president of MIT; he was being considered for the presidency of Rockefeller University. Why was this controversy over the *Cell* paper continuing?

■

Baltimore is described by friends and critics alike as arrogant and intimidating, as well as charming, earthy, encouraging to younger scientists—and, of course, brilliant. "Baltimore believed he was better than other men," one colleague said, adding: "He has reason to."

Dr. Phillip Sharp, head of MIT's Center for Cancer Research, says the center "was clearly built because of the leadership of David Baltimore and Salvador Luria. . . . He has made himself a leader in building things." And he believes that most of the talented scientists there were recruited by Baltimore.

"I grew up as a scientist at MIT in the environment of David Baltimore," says Sharp, who came to MIT in 1974. It would not surprise anyone later when Sharp came to Baltimore's aid.

"He was one of those people I never questioned," recalled Dr. Norton Zinder, a molecular biologist who was on Baltimore's thesis committee at Rockefeller University and who came to Baltimore's defense when the controversy broke.

Baltimore was a native New Yorker, born in 1938 in New York Hospital, just yards from Rockefeller University. His father and mother encouraged their talented son's interest in science. In 1952, at the age of fourteen, David attended a science workshop for high school students at the Jackson Laboratory in Bar Harbor, Maine. At the Jackson Lab, he met a seventeen-year-old Philadelphian, Howard Temin, who thought David was a natural leader even then; in a group of unusually smart kids, David stood out. "The others deferred to him," Temin recalled. That was a special summer for Baltimore—at home, he wasn't socially adept, and his smarts got in the way with other kids; but at the Jackson Lab, he was with others like himself: young people with a serious intellectual commitment. He spent long hours in the lab working on experiments. There were also picnics, swimming, and the general horseplay that occurs when kids live together. "After Jackson, I was more or less hooked," Baltimore remembered fondly. "I could spend a lifetime discovering."

Baltimore went on to Swarthmore, where he earned his bachelor's degree. He then went on to do graduate work in science at

MIT and graduate research in molecular biology in the Rockefeller lab of James Darnell, his future textbook co-author. Baltimore earned his doctorate at Rockefeller in 1964; his thesis showed how a particular mouse virus, related to poliovirus, inhibits an enzyme that acts on DNA to make its own RNA. Firmly ensconced in the world of scientific research, Baltimore worked at the Salk Institute and then returned to MIT in 1968 to teach and to do his own experiments. His goal was simple: He wanted to understand the dynamic processes that allow living beings to function—how viruses grow, what causes cancer, and how the human body fights off disease. He would eventually publish more than four hundred scientific papers.

In 1970, working on a parallel track to Howard Temin—who had also stuck with science and gone on to the McArdle Memorial Laboratory of the University of Wisconsin in Madison—Baltimore discovered a fundamental property of cancer-causing viruses. Baltimore and Temin independently discovered an enzyme, later named reverse transcriptase, which excited scientists around the world and turned science's "central dogma" upside down.

According to that central dogma, genetic—hereditary—information flows from DNA (deoxyribonucleic acid) to the messenger molecule RNA (ribonucleic acid) and then to the proteins that give cells their shape and function. Scientists believed that this process never worked the other way around. However, the enzyme discovered by Baltimore and Temin reverses the flow of genetic information.

Although the results of basic research often languish without real-world application as scientists search for all of the pieces of a particular puzzle, the Baltimore-Temin discovery had an enormous impact on biotechnology and the ongoing search for cures for cancer and other diseases. Reverse transcriptase was found in a group of viruses that cause acquired immunodeficiency syndrome—AIDS, the modern-day plague itself; they have also been linked to cancer and hepatitis. Baltimore would later explain to Representative Dingell that the discovery "was the reason that when the AIDS epidemic struck, the scientific community could so rapidly find and characterize the virus that causes that disease. The initial assay for the AIDS virus was based on my 1970 discovery."

Engineered retroviruses have become a key tool in biological research, as scientists found they could be good vehicles for gene transplants in gene therapy.

Baltimore continued his work in seeking out the causes of cancer and how the immune system works. But he also stressed in 1974 that the fight against cancer could be better met by eliminating industrial pollutants or removing carcinogens from people's diets.

The MIT scientist became outspoken on the social implications of gene splicing and biological warfare. There were great fears in the early 1970s—and concerns still exist today—that genetic engineering could let loose monstrous genes into the human genetic pool, or that engineering could be used to change the course of a person's life from its embryonic beginnings. In early 1975, Baltimore was tapped by Dr. Paul Berg, an eminent scientist from Stanford University, to be one of the sponsors of a conference on reducing potential risks from genetics engineering research.

Later that year, Baltimore made plans to move from MIT to New York to begin a sabbatical year of research at Rockefeller. Just having returned from a visit to the Soviet Union with his mentor, James Darnell, Baltimore was in New York City on October 16, 1975, to visit his father, who was in Mount Sinai Hospital following a heart attack, and to organize his move to Rockefeller. The telephone rang at 7:30 A.M.; it was Baltimore's wife, the scientist Dr. Alice Huang, calling from a conference in Copenhagen, where she had heard the news:

Baltimore, who was only thirty-seven years old, Temin, and Italian-born scientist Renato Dulbecco had won the Nobel Prize for their "discoveries of the interaction between tumor viruses and the genetic material of the cell." Baltimore and Temin had won science's most prestigious award particularly for reverse transcriptase. A *New York Times* editorial the next day said that Temin and Baltimore had "revolutionized the entire understanding of the genetic process." Dulbecco, sixty-one and a naturalized U.S. citizen in whose lab both scientists had once worked, was honored for his development of lab techniques that helped scientists study the molecular biology of animal viruses.

Baltimore tried calling his father at the hospital to give him the good news, but the telephones there had been turned off and television's *Today Show* beat him to the punch. While the elder

Baltimore was very frail, he was nonetheless able to go to Stockholm in December that year to see his son and ten other laureates receive their award from King Carl XVI Gustaf. It was the seventy-fifth anniversary of the awards, and three thousand people, including seventy earlier winners, attended the ceremony. *The New York Times* called it a "glittering" affair. (This was the year that Soviet physicist Andrei Sakharov was awarded the Peace Prize but was refused a visa by his government to accept the award. Baltimore was one of thirty-three scientists who signed a cable calling on Soviet president Nikolai V. Podgorny to allow Sakharov to accept the prize.)

Baltimore had been a vocal opponent of the Vietnam War, and in December 1981, he was a member of a delegation of four scientists appointed by Pope John Paul II to talk to President Ronald Reagan about the human consequences of a nuclear war and the need to bar the use of horrific weapons. Baltimore assured the president, however, that they were not advocating unilateral disarmament or arguing against nuclear power.

His prestige as a scientist and a social activist helped bring top-notch personnel to MIT, which suffered a bit of an inferiority complex with respect to its Cambridge cousin, Harvard. He took the lead in arguing for more resources for biotechnology research at MIT, and then for the creation of the Whitehead Institute. In both efforts, he encountered much resistance, but he fought and won people over. He really wanted the Whitehead Institute, which he could build from the bottom up and which could take a preeminent role in molecular biology research in the nation and the world.

David Baltimore became the first director of the Whitehead Institute; under his stewardship, it grew into a facility with about three hundred scientists and a $15 million budget. The institute's Kendall Square building opened in 1984. Baltimore's own lab was in the process of moving from its MIT quarters to Whitehead during the early stages of his collaboration with Thereza Imanishi-Kari. The lab moved over in 1984.

Although he ran the institute, Baltimore was also responsible for three lab groups of almost three dozen scientists. He continued to provide a leadership role in the scientific community, chairing the joint study on AIDS of the National Academy of Science and the Institute of Medicine; he was a leading spokesman for increased federal funds for AIDS research and prevention. He would later take a key role in opposing plans for the Human Genome Project, a massive national effort to map the entire gene structure of a human being, although he eventually dropped his opposition and actually chaired a private organizational meeting on the project in Reston, Virginia, at NIH Director James Wyngaarden's request.

■

The allegations by Imanishi-Kari's former postdoc, Margot O'Toole, were getting in Baltimore's way, and raising controversies where they shouldn't exist. Even before the Dingell hearing, Baltimore was getting letters and telephone calls from his colleagues and scientists around the country wondering about the case because they had heard about it from those two nervy NIH scientists, Walter Stewart and Ned Feder. What was the story? they wondered. Should they respond to Stewart and Feder? What about O'Toole?

One such letter that Baltimore had to respond to was from geneticists Leonard and Leonore Herzenberg of the Stanford University School of Medicine. They noted in a November 6, 1987, letter that they had received the Stewart-Feder manuscript and wanted to discuss it with Baltimore before replying. They were conducting similar studies and believed that O'Toole was basically correct in raising questions about the Bet-1 and whether the experiments had actually shown, not the idiotypic mimicry described, but "double producers"—cells that were producing both transgenes and endogenous genes.

The two Stanford researchers had had nothing but trouble with Bet-1: "We can easily understand how one could be honestly misled and believe that reliability had been achieved. . . . This does not appear to us to be in any sense a case of data misrepresentation although we think it's very likely that it is a case of data misinterpretation."

The Herzenbergs said that O'Toole's report to Herman Eisen was "well presented and quite reasonable." They asked how O'Toole got the data. "In essence, if Margot has a personal stake in the data and if her arguments really have merit . . . wouldn't the easiest way out of this controversy be for her to publish her views in a letter to *Cell* with you as co-author? Since she seems to be the first to recognize the possibility that transgenics have double-producing B cells, we would find it more comfortable to be able to refer to such a letter in our upcoming paper on the double producers. This would make our argument easier since the lack of double-producing hybridomas has been . . . most difficult to reconcile with our current findings."

The Herzenbergs added that, if O'Toole had a legitimate claim to relevant data, "we could consider inviting her out here to participate in the production of the new hybridomas and thus in the publication of the new results."

Like Baltimore's friends and colleagues, the Herzenbergs noted that their questions did not indicate that they believed there had been any scientific dishonesty. They tactfully wrote: "We certainly don't mean to imply that you've dealt badly with [the matter], nor do we mean to imply anything negative about Thereza's actions."

Baltimore replied on November 12 that the Stewart-Feder manuscript was based on data that he assumed were supplied by O'Toole; he did not know if any of the data were hers. "I am not sure what Margot was doing with Thereza; I do not believe that I ever met her when she was in Thereza's lab," Baltimore said, adding that he was glad that they viewed the dispute as the "usual scientific uncertainties."

Twelve days later, the Herzenbergs wrote that on the basis of Baltimore's response as well as information from Henry Wortis, they didn't believe there was any reason to invite O'Toole out to California to work on the new experiments.

As soon as the *Boston Globe* article appeared and the first Dingell hearing was held, Baltimore knew he had to go on the offensive, or everything he had worked for in his life would be overwhelmed by a ruined reputation. He was a builder of institutions, a leader of the scientific community, and he would not be destroyed by the allegations of a postdoc about, not his own work, but that of one of his collaborators. His efforts to calm O'Toole down had failed and he couldn't make Stewart and Feder go away.

At the beginning of the controversy Baltimore had depended for advice and help on an administrator in his office, on a part-time public-relations aide who worked for the Whitehead Institute, and on his own lawyer, Normand F. Smith III, a tax and estate-planning specialist. Because of the short notice of Dingell's inquiry, Smith quickly contacted a Washington, D.C., lawyer, Marc Ginsburg, a former aide to Democratic senator Edward Kennedy of Massachusetts, with whom he had worked before, and asked him to sit in on the congressional hearing in April. Soon afterward, the team decided that Baltimore needed a powerhouse law firm with the political savvy to guide him through Dingell's inquiry. Through a friend, the scientist found the D.C. office of Akin, Gump, Strauss, Hauer & Feld. Akin, Gump was home to Robert Strauss, the former Democratic National party chairman and ambassador to the Soviet Union. The firm made its name through its political connections.

"We felt he needed to be represented by a Washington firm," Smith told the *Legal Times*, a weekly newspaper that covers law and lobbying in Washington. "Up in Boston, we're very provincial. When you want to know what's going on in Washington, you hire a lawyer in Washington."

Baltimore later agreed. "I didn't know anything about John Dingell then. I was looking for someone to give me advice."

The Akin, Gump team, paid for by the Whitehead Institute, included Daniel Joseph, a former special assistant to the general counsel of the U.S. Department of Transportation; Daniel Spiegel, a former aide to the late Hubert Humphrey and to Senator Alan Cranston; and associate Marydale DeBor. Also on the team was partner and chief lobbyist Joel Jankowsky, once an assistant to Carl Albert, the former speaker of the U.S. House of Representatives.

The Washington lawyers came up with plans to attack the congressional inquiry. It was clear to them and Baltimore that he was the target, even though his lab's work was not in question, because of his name. He was the big-shot scientist, and Dingell went after big-shots. So, with the agreement of his co-authors, who were more than happy to let him take the lead, Baltimore became the spokesman. One of the first things he did was write a nine-page letter that went out May 19, 1988, on Whitehead Institute stationery, to four hundred scientists around the country. Baltimore said it was his own idea to write the letter: "I felt I had to communicate with my peers. People like John Dingell control the media. And I had a lot to say. I had a very intense desire to do that."

The "Dear Colleague" letter was a stinging indictment of Stewart and Feder, Peter Stockton, and the news media. In ominous tones, Baltimore warned of threats to the scientific community. He was defending not only his own reputation but American science itself.

"A small group of outsiders, in the name of redressing an imagined wrong, would use this once-small, normal scientific dispute to catalyze the introduction of new laws and regulations that I believe could cripple American science," Baltimore wrote.

He described the experiment's background, saying that he and Imanishi-Kari had different interpretations of their findings. Although the paper's summary presented idiotypic mimicry as its central conclusion, Baltimore said he "felt the explanation probably lay elsewhere," perhaps in gene conversion. He noted that a paper he later co-authored with the Herzenbergs suggested that "much of the idiotypic reactivity is a consequence of the high expression of the Ly-1 B cells."

Baltimore explained O'Toole's questions as a matter of alternative interpretation of the data, and said: "I do not believe that the question is at all closed and there may well yet be other surprises and other interpretations."

The capability of Bet-1 was indeed overstated, Baltimore wrote, but he added that some preparations of the reagent were better than others. He indicated that he and Herman Eisen of MIT at one point doubted the specificity of Bet-1 at all, but he blamed that on a misunderstanding due to the fact that English is not Imanishi-Kari's native language. (This incident would later become a major embarrassment to Baltimore at the following year's congressional hearing.)

Baltimore argued that the two university reviews led him to believe that it was "unlikely" that O'Toole's concerns were valid and he said only new experiments with new techniques would be helpful.

"There never has been any effort by me to discourage Dr. O'Toole from pursuing her questions, nor have I done anything to affect her career," Baltimore wrote.

Baltimore's supporters have since tried to reconcile his account with O'Toole's. They believe that he may not have realized that what he saw as forceful advocacy—the promise to respond to any letter O'Toole wrote to a journal—she may have seen as intimidation. Perhaps, also, Baltimore did not recognize his letter to the Herzenbergs as an action that would discourage them from taking on O'Toole at Stanford. (Leonore Herzenberg later said that Baltimore had indeed discouraged her from bringing O'Toole out to do further research with them, and that she regretted "that I listened. . . . We would have proved right there that there were

quests for documents. Dingell's subcommittee had issued broad subpoenas for the original data and documents relating to the experiments. But unbeknownst to Baltimore and his co-authors, Dingell had also called in the Secret Service to take a look at the original notebooks from Imanishi-Kari's lab. It would be difficult, perhaps impossible, for Dingell to argue the complex science, but the Secret Service could study the original papers and determine if there was forensic evidence that some of the experiments were not done when claimed.

Baltimore did know, however, that turning over Imanishi-Kari's lab records would be somewhat difficult. Formally called notebooks, the records were actually a hodgepodge of paper and radioactivity-counter tapes.

SEVEN

The Accused

■

The cigarette butts were piling up. She was a two-pack-a-day smoker, and there was no way she was going to be able to stop any time soon. Too, Imanishi-Kari was confused; she simply didn't understand why the "Department of Energy and Commerce" was interested in her. "This is so weird," she thought when the Tufts provost informed her that there would be a congressional hearing about her paper in *Cell.* A Brazilian citizen, she says the only thing she ever knew about U.S. congressional hearings was the infamous McCarthy hearings.

Following the disturbing meetings with Henry Wortis—she had worried for a while that her Tufts appointment was in jeopardy—and Herman Eisen, she thought all the problems with Margot had been resolved. But then, while rushing to get out of the lab for some last-minute Christmas 1986 shopping she was hit by a telephone call from some NIH scientist named Walter Stewart asking for the data behind the *Cell* paper. She didn't recall hearing about the notorious seventeen pages until the Stewart-Feder manuscript actually was written and circulated. But David Baltimore had agreed to take the lead role in dealing with this nuisance; and Thereza was only too glad to let him do so. Perhaps David's preeminence would shut this challenge down; certainly he had more power to do so than she, an assistant professor at Tufts and not even an American citizen.

She thought he had managed to resolve the Stewart problem. And now, another year had gone by and reporters were calling in

April about charges by Margot and that horrid Charlie Maplethorpe. Maplethorpe's accusation of fraud came as a shock to her; he hadn't been involved in the experiments and while he had always been argumentative, she didn't recall any suggestions from him that she had made things up. Imanishi-Kari managed to avoid talking to the reporters, but it was clear from the May congressional hearing that those two former postdocs were going to haunt her. Imanishi-Kari compared hearing of the accusations, made in the very public congressional proceeding, to being raped: "It violates you tremendously." At least David was still backing her. But she didn't understand the fuss. Nothing she had read as a child in Brazil about the great Madame Curie, whose life story had inspired her to enter science, had prepared her for these challenges to her work.

■

Thereza Imanishi was born on December 14, 1943, in the Brazilian town of Indaiatuba, where her Japanese parents had a farm. Her father immigrated to Brazil with his parents when he was three years old; her mother was second-generation. When Thereza was in her teens, her father set up a company to transport other Japanese farmers' produce to a cooperative. A long way from Japan, they nevertheless sought to lead a traditional life and struggled with their three daughters' desire to move away from home and go to college. The two sons were not a problem, but the girls—Thereza was the middle daughter—were determined to live in the most modern way.

"My parents wanted us to be housewives or teachers, so you could have a family," Thereza remembered. But when a teenaged Thereza read a book about Madame Curie, she knew she would never be happy marrying and staying in Indaiatuba. "The idea of doing research, the idea to do something that you could spend day and night doing . . . , it would be fascinating," she recalled. "I wanted to do something so exciting that I could work day and night."

Thereza Imanishi earned her undergraduate degree in 1967 at the University of São Paulo, and then won a scholarship to do graduate work at the University of Kyoto in Japan. Her grandfather wanted her to see the land of her ancestors. But while she earned the equivalent of a master's in developmental biology at Kyoto and thought the country was beautiful, she had no interest in staying any longer than necessary. It was a time of student unrest and the political hubbub was getting in the way of her education. "When I was there in 1968, 1969, there was a student, how do you say, uprising, so the university was closed. . . . I did not want to throw stones at the police, not that I was not aware. . . . I wanted to do something but not that way."

Imanishi moved on to Finland to work on her doctorate in immunogenetics at the University of Helsinki. The young scientist was attracted by the lure of Scandinavia's socially progressive reputation, as well as by the work on antibodies being done by scientist Ole Makela.

While Imanishi was starting on her life's work in immunology and batting around the idea that perhaps she should be doing something more immediately useful, such as medicine, her younger sister, Tochi, had become quite ill. At the time the family only knew that Tochi had serious kidney problems. Tochi recovered enough to go off to England; she became an architect and returned to São Paulo to work as a city planner. Meanwhile, in 1972, Thereza met and married a young Finnish architect, Markku Tapani Kari. Three years later, they had a daughter, Elina. Imanishi-Kari moved on to Germany, where she worked in the laboratory of Klaus Rajewsky at the Institute for Genetics of the University of Cologne.

Imanishi-Kari's life seemed to consist of major highs and lows, and in 1981 the roller-coaster pattern recurred with particular intensity. Despite her initial recovery, Tochi learned she was much more seriously ill than she had thought: She had lupus, an autoimmune disease, which flared up after her second child was born. Tochi died of complications from lupus in 1981. Also that year, Thereza and Markku separated.

Then, completely unexpectedly, a letter arrived from Herman Eisen of the Massachusetts Institute of Technology: MIT was looking for someone with a background in cellular immunology. Would Imanishi-Kari consider accepting a position as an assistant professor at the MIT Center for Cancer Research?

At MIT, Thereza might be able to collaborate more directly with David Baltimore, who had done some projects with her boss, Rajewsky, and who needed an immunologist to analyze the antibodies for experiments he was considering. Imanishi-Kari had known and worked with Baltimore since 1979. The idiotype system that Baltimore wanted to test had been defined first by her, in Finland, and she had studied it further in Germany.

Imanishi-Kari joined the MIT faculty as an assistant professor in 1981, but she did not begin working with Baltimore until she had had some time to establish herself independently.

In 1983 and 1984, Baltimore began a collaboration with Imanishi-Kari's lab because the gene that he wanted to study was related to the group of antibodies she had been studying for years. Baltimore's postdoc David Weaver worked in her lab for a while, learning about fusion and subcloning techniques.

■

Life in the Imanishi-Kari lab was anything but sedate. Notes on experiments were written on odd pieces of paper stuck in manila folders, in notebooks, and on pads, and kept in haphazard fashion in and on top of desk drawers along with cut-up print-outs of radioactive counts. A heavy smoker even in the lab, Imanishi-Kari easily lost her temper, and was not shy about telling people what she thought of them. "Thereza is very intense . . . [with] high expectations. . . . She's a very sweet person, but she calls them as she sees them. She calls me a twit all the time," said Joan Press, a scientist at Brandeis and a friend of Imanishi-Kari's. Thereza dressed in shorts and seemed to like to shock the undergraduates and others around her with provocative language in her charmingly accented voice. "Maybe I do have a loud personality," she admitted.

Charlie Maplethorpe, in particular, drove Thereza nuts; she would often shout at this nervy graduate student. Imanishi-Kari thought he was very smart, but Maplethorpe was constantly questioning her authority, arguing against her proposals for experiments, and generally bitching about her. He would complain that she was secretive and would not share information about experiments with him. Basically, he was a major irritant to her; he showed her no respect. But despite her own brash manner, she did not have the guts to kick him out.

Imanishi-Kari also had a "difficult" relationship with Margot O'Toole, even though they had mutual friends. Margot was very protective of her place in the lab and did not want anyone doing experiments similar to her own. When they disagreed on how to handle a particular experiment, Margot acted infuriatingly self-assured—she was right, Thereza was wrong. Although she didn't think her relationship with Margot was as bad as that with Charlie, Thereza again felt that she was being put down. "This is my postdoc telling me what to do in my lab," Imanishi-Kari said.

But most of all, she thought that O'Toole was of two minds about being a full-time scientist while raising a family. Imanishi-Kari says she pushed O'Toole to apply for her own grants because she, Imanishi-Kari, would be leaving in June 1986 for a new post at Tufts; but, she contended, Margot resisted doing grant applications because she wasn't sure that she wanted to work outside the home. "I think she really wanted to have more children," Imanishi-Kari said. "If you want to have more kids, it's hard to be independent. You have to write grants."

But despite their flare-ups, Imanishi-Kari was pleased about the results of some experiments that O'Toole had done with Moema Reis on adoptive cell transfers. The experiments provided the beginning of evidence for the mechanism behind the theory that cellular interactions could be the reason that the mouse gene was producing a foreign idiotype. Imanishi-Kari presented the results on O'Toole's behalf at the New England Immunology Conference in October 1985. The results grabbed Baltimore's attention, and he and Weaver met with Imanishi-Kari and O'Toole later that year to discuss O'Toole's work. They decided to mention that work in the *Cell* paper on idiotypic mimicry, which they were then preparing for publication, and they asked O'Toole to critique a draft of the paper. In fact, Imanishi-Kari said it was these results of O'Toole's, never published, that tipped the central conclusion of the *Cell* paper toward idiotypic mimicry. "The reason [the paper] favors the network theory was because of Margot's results. How can I ask a postdoc to prove the hypothesis if I had falsified, unless I'm crazy?" Imanishi-Kari asked.

Imanishi-Kari was looking forward to the spring publication of the paper and to her planned move to Tufts, where she had more friends. Despite her collaboration with Baltimore's lab, she had found it difficult to work with other scientists at MIT. Although she did not think she was being secretive within her own lab, as Maplethorpe charged, Imanishi-Kari said she tried to keep a lid on some of her experiments to avoid provoking the competitive jealousy of some of the other scientists. (She later said all her records were available in the lab.) She actually started looking for another job just a couple of years after joining MIT. She said she knew she wasn't going to get tenure: Dean Brown and others were pushing her to publish more, and she was refusing to do so. "I'm not asking you to give me tenure," she told Brown once. "I will write when I have data." Senior scientists at MIT later said that while Imanishi-Kari was certainly good enough to have been hired in the first place, she never attained the independent stature that they looked for in a tenured professor.

Imanishi-Kari thought Moema Reis was a pleasure to work with compared to her other postdocs. Reis worked hard, she produced, she didn't cause a fuss. Her expertise was in mouse genetics and establishing new lines of mice. The results of Reis's fusion experiments were presented in Table 2 of the *Cell* paper. But relations with Margot became more tense after Moema returned to Brazil and O'Toole could not repeat her own results. Someone had to be responsible for the mice, so Imanishi-Kari asked O'Toole to join her in breeding and typing the rodents. But after more arguing over O'Toole's continuing failure to duplicate her earlier experiments, Imanishi-Kari told her to just take care of the mice. Thereza was too angry and tired to continue fighting over the experiments.

■

Thereza had been bone-tired for some time. She had gotten puffy and found some strange scratchlike marks that she couldn't remember getting. A doctor at the MIT health clinic said it was a no-name rash and gave her some cream. But then after returning from a Lake Tahoe immunology convention in early 1986, Imanishi-Kari found she couldn't open a door, her hands had swelled up so badly. She had trouble walking; her joints hurt. Breathing became a little harder, but she put that down to her heavy smoking. Worried, however, she told her friends at Tufts; a fellow immunologist thought her symptoms sounded like a typical autoimmune disease. A specialist then confirmed it: Imanishi-Kari had lupus—not the virulent form that killed her sister Tochi but still cause for worry.

Her doctors tried a series of drugs, but finally a malaria remedy, and ibuprofen for the pain, got her into fairly good shape. Clearly 1986 was not a good year. Besides her illness, she was finally divorced from her architect husband; even though they had been separated for some time, the emotional impact of the divorce hit hard.

Thereza began the process of winding down her MIT lab for the move to Tufts. Then she got a call from Henry Wortis, the man who was recruiting her for the Tufts pathology department, asking if she could come by with her notebooks. Imanishi-Kari remembered going over all of the data except for some X rays. A second meeting, which included O'Toole, was focused on those X rays and other data. They discussed alternate explanations for the results, Imanishi-Kari recalled. But she insisted that she didn't call O'Toole and scream at her or threaten legal action. And she didn't remember crying at the meeting when she asked if the other scientists didn't believe her. "I don't remember. But I was very hurt that anyone would question my integrity. It is so hard," she said, looking away and crying.

Imanishi-Kari insisted there was never any talk of a retraction at the meetings. She did recall, however, rejecting O'Toole's outstretched hand at the end of the difficult meeting. "You caused so much trouble," Imanishi-Kari said. "If you have scientific issues, come to me."

That was a fateful decision, Imanishi-Kari later believed. She wounded O'Toole's pride and angered her, angered her enough to continue her challenge until she lost control of it. "In retrospect," Imanishi-Kari said, "my mistake was I should have shaken her hand. I couldn't do something I didn't believe in."

The challenge did not stop with the review by Wortis, Huber, and Woodland. Herman Eisen, the MIT professor whose letter brought Imanishi-Kari to Cambridge, called in June to tell her that O'Toole was very emotional and wanted to meet to discuss the paper again. Imanishi-Kari thought it strange that Eisen did not want to see her data.

She also contended that there was no talk of a possible retraction at the Eisen meeting with Baltimore because of the problems with the Bet-1 reagent. Imanishi-Kari never thought that a correction was necessary simply because the paper overstated the reagent's capability.

After the meeting, she ran into Eisen in the hallways of MIT on occasion, when she returned from Tufts to pick up her hybridomas and papers. He kept coming back to some point or other that bothered him about the *Cell* paper, but she didn't recall what they discussed. Imanishi-Kari said, however, that she would never have told him that the Bet-1 reagent had never worked as claimed.

Eisen also never told her about a letter that Baltimore wrote to him, blasting Imanishi-Kari for allegedly saying that the Bet-1 reagent did not work. She would not find out about Baltimore's criticism until shortly before the 1989 congressional hearing. For now, in 1988, she found that she had to continue her fight for her professional career and reputation. But she didn't have the financial resources of a David Baltimore; a friend directed her to Bruce Singal, a former federal prosecutor and now a partner in the Boston office of the small Holyoke, Massachusetts, law firm of Ferriter, Scobbo, Sikora, Caruso & Rodophele. Singal would handle the case virtually pro bono.

It was now June. The NIH investigators, as well as the congressional staffers, wanted to talk to Imanishi-Kari, but another fateful event, more important than the refusal to shake O'Toole's hand, occurred before they came to Cambridge. She had talked with Baltimore and his lawyer about the situation, including how she should deliver her lab records to the government officials; the subcommittee had issued a June 30 subpoena for all her documents. Baltimore believed she should just hand over boxes of her papers as they were: a sheet of paper here, a counter tape there. Normand Smith, however, recommended cataloguing her records and making them "as comprehensible as possible." Imanishi-Kari agreed with Smith. "He didn't force me."

But while Imanishi-Kari, Smith, and Baltimore said that all that was done was to organize existing data, congressional and NIH investigators would put another spin on it.

The Secret Service

■

From David Baltimore's letter to four hundred scientists around the country and the reaction of the scientific community to the congressional hearings, John Dingell and his investigators quickly recognized that Baltimore and Thereza Imanishi-Kari were going to dig in their heels and throw a lot of science at them. Dingell knew that it would be all but impossible to fight with the distinguished Nobel laureate on the basis of the science—which wasn't his point, anyway. And once Baltimore took up Imanishi-Kari's fight, it would not be easy for either man to step back.

"If Baltimore hadn't raised as much of a stink, I don't think Dingell would have remained interested," recalled Michael Barrett, who was then staff director of the subcommittee. "The scientific community was really very arrogant. Dingell felt nobody was above question."

But how could the subcommittee move the investigation forward on terms it could defend?

Ever since Dingell's aides got their hands on the original lab notebooks, they had noticed little date changes and marks on the radioactivity-counter tapes that had made them suspicious. They wondered if there was some way they could independently date the pages of the experimental data. Before they went further, however, they had to lay the groundwork. On March 7, 1989, Dingell wrote to Imanishi-Kari asking her to certify that the lab books she had provided under subpoena "are indeed the original lab books that contain the original data, were created contemporaneously with the scientific experiments, and were prepared prior to the submission of the *Cell* paper."

The response came from Imanishi-Kari's attorney, Bruce Singal, who on March 27 gave a very qualified, lawyerly reply. He noted that the notebooks contained all of the recorded data from the paper along with "confirmatory and ancillary data" and experiments from Imanishi-Kari's continuing research. Not all of the data were recorded personally by her and "furthermore, the notebooks were kept in the laboratory and accessible to anyone in the lab." Singal also noted that the binders may have been changed "but the pages contained in the notebooks are those on which the original data were transcribed."

"Contemporaneous" had to be defined. Strictly speaking, Singal said, only the counter tapes and the X rays were created at the same time as the experiments; the handwritten entries necessarily came after the conclusion of the work. Singal pointed out several notations, including some made by O'Toole during the Tufts review, that postdated the paper's publication. But, he said, "construing the word 'contemporaneous' in the broader sense, it is fair to say that all of the data presented in the *Cell* paper were assembled contemporaneously with the scientific experiments and placed in the notebooks in a timely fashion. . . . The data presented in the *Cell* paper and recorded in the notebooks were prepared prior to the submission of the *Cell* paper."

Dingell and his aides believed that if they could show that some of the substantive work was recorded out of order they would have evidence of misrepresentation or even fraud. Vulnerable on the scientific front, Dingell the hunter would resort to bigger guns, which happened to belong to the Secret Service.

The image of Secret Service agents is pretty well set in the public's imagination. They are the brave ones willing to throw themselves in front of the President of the United States to stop an assassination attempt without a thought for their own well-being. They're also the guys with the hearing pieces in their ears, who look at everybody but the President when he's gripping and grinning with the public. But the Secret Service does much more than protect the President.

The agency is also expert at forensic investigations of written documents, and it maintains the International Ink Library, the world's largest collection of writing-ink formulas—more than seven thousand samples. The vials of inks and the binders of representative scribbles are kept in unprepossessing metal cabinets.

Using its library of samples, some rather unsophisticated detection devices, and some gee-whiz technology, the Secret Service is responsible for investigating currency counterfeiting, letters threatening the President, Food Stamp fakery, and Medicaid cheats. Credit-card fraud also comes under its aegis; the thinking is that the card is a bank instrument, so fraud could undermine the banking system and the nation's economy. Agents can even determine where the plastic in a particular card comes from.

Because of their expertise and ink library, the agents at the Questioned Document Branch are called by other agencies and governments for help in cases where documents are involved. Larry Stewart, the lead document examiner for the instrumental analysis section, helped the Canadian government crack two murder cases: He blew the murderers' alibis by analyzing diaries and checking the ink and paper of other records. The Secret Service also investigated documents for trials of Nazis in Hungary, Canada, and Australia.

John Hargett, chief of the Questioned Document Branch and the agency's senior document examiner, took the call from Peter Stockton in late August. Could he and Walter Stewart come over and talk about a possible problem? Stockton and Stewart left the confines of Capitol Hill for the agency's downtown office, one of D.C.'s ubiquitous marble buildings a couple of blocks from the gingerbread Old Executive Office Building and the White House. Hargett's floor had a guard behind a window, a sign explaining the correct way to wear one's security badge, a laboratory neater than the average academic lab, and a number of locked doors with special security devices. Hargett's office, though modest in size, was actually fairly elegant for someone who wasn't a political appointee: a large mahogany desk; a mauve-, blue-, and gray-striped sofa; and mauve carpet. It was definitely fancier than the environs Stockton and Stewart were used to on the Hill or the NIH campus.

Hargett laid out what the Secret Service could do in terms of ink and paper analysis. As it turned out, the agents' ability to find impressions on the data sheets that were imperceptible to the naked eye would outshine the notable ink library. Hargett told Peter and Walter that it would be no problem to add the project to his branch's other work—his agents had had assignments from other congressional committees in the past—but he would need a formal request. He got that in late August, along with two Imanishi-Kari notebooks—I-1, which was the 1984 data notebook for the lab, and I-3, the 1986 notebook. Some twenty-six notebooks from various MIT labs would eventually be studied.

Hargett had just one condition: Don't tell us about the scientific research involved in the notebooks, don't explain how the experiments are supposed to work. It was easier for the agents to be objective about their findings if they had to deal only with the ink and the paper. "We had no idea about the research conducted at the time," Hargett recalled three years later, laughing, "and we still have no idea about the research.

"There were times they'd feed us information and we'd disregard it. Our work should be independent from the research."

Though personable types, Hargett—a rare D.C. native—and Larry Stewart were clearly buttoned-down government employees, compared to the free-wheeling Dingell aides. Anyone would be hard-pressed to recall the last time Stockton ever wore a suit. The congressional staffers' irreverent speech and chaotic offices made an impression on the Secret Service agents, but they got along fine.

Perhaps they felt a bit of an affinity for each other: Their non–Ivy League credentials were up against a Nobel laureate and the Cambridge crowd. Referring to Bruce Chafin and himself, Stockton liked to say: "We're the land-grant guys."

Hargett's academic credentials were even less polished. He was from the "old school": Although he attended many schools, he never earned a degree. But by 1988, he had seventeen years of questioned-document expertise in the federal government and five years' work in private practice. Larry Stewart, who headed a professional association panel that developed standards for ink analysis, had earned an associate of arts degree from Florida Technological University, a bachelor of science degree in forensic science from the University of Central Florida, and a master of forensic sciences from Antioch University. After three years at the Bureau of Alcohol, Tobacco and Firearms, he joined the Secret Service in 1982.

Steve Herzog, the agent who examined the *Cell* paper's photograph of X rays of RNA molecules, had twenty-nine years of professional experience and a bachelor of science degree from Bob Jones University in South Carolina.

Speaking of the Dingell aides, Hargett said "They had a little more leeway from other people. They were a little rough around the edges. That's just the way they have to be to get things done. Within a red-tape world, you have to be a little boisterous."

Hargett had several staffers first run one thousand pages of the lab records through the British-made Electrostatic Detection Apparatus (ESDA), a simple device originally designed to reveal fingerprints. A document is first placed over a pan of water, to impart some humidity to it, and then put on top of a metal mesh. A sheet of plastic paper—Hargett likes to say "plastic membrane," but it looks like Saran Wrap—is stretched over the top; the ESDA sucks the document and the plastic tight onto the metal. The metal plate is charged and a toner carried on small glass beads is poured over the plastic. The toner clings to the plastic over the charged areas. When the plastic is peeled off the document, it carries with it a copy of any indentations or impressions from the page. The ESDA can even uncover indentations made by a felt-tip pen.

Over the next eight months or so, Hargett and Stewart painstakingly compared one thousand individual laminated sheets to see if any of the impressions matched up with the writing on other pages of the lab records. They would spread out the papers in the branch's small conference room and grab a couple of hours or sometimes several days between other projects and patiently flip through the pages. It was almost like a free-floating jigsaw puzzle, trying to find a plastic page that would line up with an original page.

Their analysis started to take form and they were able to reconstruct the sequence in which the pages were written on and torn off a pad. The pages that matched indentations detected by the ESDA were set aside for further study, along with other pages on which the two agents could easily see obvious date changes or the moving of adhesive tape holding the radioactivity-counter tapes.

Larry Stewart then studied the tapes, which recorded the counts of the radioactive reagents and thus measured the strength of the reactions. But Stewart did not know any of this. He saw the tapes as yellow and green pieces of paper with typed information on them. With time and use, the typewriter ribbon loses intensity and must be changed; the paper roll also runs out and must be replaced. The paper, made totally of wood pulp, was of poor quality, and the dyes varied. Stewart was sure he could examine the counter tapes in the lab books and determine through the intensity of the ink and the color of the paper when the ribbon or the paper was changed.

Stewart noticed what he considered drastic changes in the tint of the paper in places where he did not expect it, unless more than two counters were being used for the experiments. In the 1984 lab book, for example, he found that the tape was dark on page 119, very light on page 122, and dark again on page 123. But the agents were not quite sure how to test the counter tapes further.

Larry Stewart found a number of dates that were altered throughout the 1984 notebook; these appeared to him to be innocent changes, but three particular changes stood out to him and Hargett. Looking through a microscope at page 96, Stewart saw an obliteration in the area of the "9" in the year "1985." He then moved to an instrument called a visible microspectrophotometer, which basically allows a closer inspection without taking a sample of the ink. Using that instrument, Stewart found that two different inks were involved in making the date: The first ink made a 10 and the slash and part of the messed-up area, while the second ink made the 1, the first slash and the 85 of the year, and part of the overwriting on the 19th.

Later, analyzing the indentations in the paper, the agents determined that the date originally written on page 96, as well as on pages 97 and 98, was "10/12," or October 12. Indentations from page 96 were found on page 36, dated "10/13/84." The agents were not sure of the meaning of what they had found, but they believed that Imanishi-Kari had gone to some lengths to change "October 12, 1984" to "January 10, 1985." (A scientist consultant for NIH would later testify that it was unlikely that the subcloning experiments described on that page could have been done in October.)

The agents also were able to determine that a number of the pages in the notebook came from pads in which the pages came glued together at one end. They were able to show that at least two different pads, A and B, were used, because of how the pages were lined and because of printing defects from the paper-manufacturing process. Also, the Pad A paper turned out to be thinner than the Pad B paper.

The 1984 notebook had no pages from Pad A, although the 1986 notebook did, along with pages from Pad B. Then, by tracking the imperfections in the pages, the agents were able to determine which papers came from close proximity to others in the two pads. For example, Stewart determined that pages 30, 41, 43, and 113 in the 1984 notebook had once been close to pages 20 through 25 in the 1986 notebook.

Stewart continued studying the pages through ink analysis, finding that the same blue ball-point ink was used on page 30 in the 1984 notebook, and then again on page 113. While that by itself would not be unusual, Stewart later found that a black ball-point pen was also used on pages 113, 41, 42, and 43 in the 1984 notebook; the same ink was used on pages 20 through 25 of the 1986 notebook. Although the ink analysis could not be conclusive, it certainly indicated to Stewart that several pages in the 1984 notebook had actually been prepared in 1986.

Going back to the ESDA analysis, the agents were able to determine the sequence in which those pages were prepared. Finding the impression of the October date from page 41 on pages 42 and 43 of the 1984 notebook linked those pages for them. Then, on page 41, they discovered indentations that matched writing on page 113. Also on page 41, they found impressions of the edges of page 113. This indicated that page 113 was on the pad directly on top of page 41.

Another page they found suspect was page 30 from the 1984 notebook. The ESDA analysis showed that page 25 of the 1986 book was prepared on the pad directly above page 30 of the 1984 book. The two pages purportedly had been prepared two years apart.

The indentation analysis also allowed the agents to see other changes in the dates. The date on page 89, for example, appeared to be 12/12, but by using the ESDA on page 91, two pages below, the agents found that the original date was 10/2.

Page 107, dated 2/7, in the 1984 book was prepared before page 31, dated 10/30; page 121, dated 5/22/85, was written on top of page 5; and page 119, dated 3/22/85, was done on top of page 7, which had an altered date of 8/22/84, page 8, dated 8/22/84, and page 9, dated 8/26/84.

What the document examiners reported to the subcommittee staff was that one page was purportedly prepared about two years later than another when in fact they were both written about the same time; another four pages were also produced around 1986, rather than on the claimed 1984 date. On the basis of the May 19, 1986, date on earlier pages, they believed the five questioned pages were produced at that time. The agents also questioned the sequence of another twenty pages.

What the agents did not know was that May 19, 1986, was several days after Margot O'Toole first raised questions with Henry Wortis, her former mentor at Tufts. According to O'Toole, Imanishi-Kari had not brought all of her data to the May 16 meeting with Wortis, Huber, and Woodland. After O'Toole argued that the panel had not seen the relevant data, a May 23 meeting was held at which Imanishi-Kari brought two more sheets of data and said she "resented" having to generate them in one week.

Hargett and Stewart were joined in the last month or so of their investigation by Steve Herzog. His job was to examine the *Cell* paper's Figure 4, an obviously composite photograph of X rays from the Northern blot analysis determining whether some cells with transgene properties had transgene or endogenous RNA.

Figure 4, produced by Baltimore's former postdoc David Weaver, who was now an assistant professor at Harvard and the Dana Farber Cancer Institute, was an important cross-check, or confirmation, of the paper's conclusions. Only one cell was expressing transgene RNA, according to the photo composite, while three others with transgene properties were actually endogenous.

However, by exposing four subpoenaed X rays for varying periods, Herzog found that another cell was showing a smudge, or band, that did not appear in Figure 4. Herzog was able to reproduce the paper's Figure 4 only by stripping and combining two of the negatives with longer exposures. The *Cell* paper had said nothing about the composite nature of the photograph, the different exposure times used for the various cells, or the existence of another band.

Although he could tell that the smudge was left out of the final photograph that was printed in the paper, Herzog did not know that the smudge was the size of the transgene's RNA. The X ray had actually shown some evidence of a double producer—a cell expressing both a transgene and an endogenous gene—even though the *Cell* authors had scored the gene only as the mouse's own.

Presentation of their findings would not be too difficult for the Secret Service agents. The subcommittee, their "client," did not ask for an extensive formal written report as is sometimes prepared for court testimony. The document examiners only needed to do an outline, prepare some blow-ups for presentation at briefings, and be ready to testify at a subcommittee hearing.

They met Representative Dingell only once before the subcommittee hearing. Hargett and Stewart joined Stockton and the other aides at a briefing in Dingell's office. "There's an air about him that hits you when you come in the door and he's sitting behind his desk," Hargett says. "Congressman Dingell runs the meeting and everybody around the room knows that he runs the meeting."

But Hargett and Stewart, who had been involved in controversial, high-stakes cases before, said they never felt pressured to come up with something that wasn't there. Said Hargett: "Having sat in Congressman Dingell's office, there's no question that's a powerful man. But we're used to dealing with people who want an answer."

NINE

The First NIH Review

■

It was the winter of 1988–1989, and the federal scientific bureaucracy in Washington was concerned about the potential of John Dingell and Congress to do mischief.

Dingell had shown no signs of letting up on his investigation into scientific misconduct in general and the David Baltimore case in particular. He planned to hold another hearing in a few months. There were legislative proposals floating around in Congress that would take the handling of scientific-misconduct cases away from the scientists. After a shaky start, the NIH panel had finally gotten off the ground, interviewing the authors of the *Cell* paper and reviewing the data; the panel's final report was being reviewed by NIH's senior staff. But the scientists in the government did not believe that the report would be enough to satisfy Dingell, conclude the controversy, and lay the legislative threats to rest. "We needed a preemptive strike," recalled Dr. James Wyngaarden, the agency's then director.

Officials of the Department of Health and Human Services, the Public Health Service, and the NIH met at the HHS under secretary's conference room "downtown"—the Hubert H. Humphrey Building, in southwest Washington—to determine what they should do. Under secretary Don M. Newman's conference table easily sat thirty people and had room for the crowd.

Robert Charrow, the deputy general counsel of HHS, wanted the department to adopt his scientific-misconduct task force's recommendations for a more formal legal structure. He wanted an investigative group clearly separated from the adjudication side, and he wanted the accused to have the right to a hearing and the right to confront witnesses before any findings, no matter how preliminary, were made. Charrow, who paid his way through law school working on laser physics at private companies, believed that scientists relied too much on collegiality to be able to handle serious investigations fairly and in a timely fashion. "When people's careers are at stake, these disputes become as bitter and violent as divorce cases," he explained.

But the government scientists did not want to lose control of misconduct investigations. Wyngaarden was on the side that believed it was vital to keep the cases in the hands of scientists, who would be able to understand complex scientific matters and the subtleties of scientific misconduct. "Most of these charges are not legal matters. I did not want to leave it to the lawyers," says Wyngaarden.

Wyngaarden's side won. HHS drafted plans to establish an office within NIH, which would handle investigations, and an office within HHS to deal with the final adjudication and mete out sanctions, if necessary. In fact, the lines were much blurrier: The first office ended up drafting reports that included very definite findings of guilt even though they were tagged as preliminary, and the accused did not have the right to confront the accusers before the case was bumped up to departmental level.

Disdainful of the decision, Charrow tagged the NIH office with the name "Office of Scientific Integrity." It was a joke: The name sounded Orwellian to him. The HHS office became known as the Office of Scientific Integrity Review. As his final gesture of opposition to a system that he didn't think would work, Charrow, as deputy general counsel, held up the establishment of the OSI until he left the government April 1; he called this inaction his "pocket veto." Charrow returned to private practice, joining the Washington, D.C., law firm of Crowell & Moring, where he represents scientists accused of misconduct.

■

The NIH report on allegations of scientific misconduct in the Baltimore–Imanishi-Kari case, however, began, under the old system, in the NIH Office of Extramural Research. And OER was moving exceedingly slowly. Around the first of the year, NIH appointed its ill-fated panel of outside scientific experts, which included Baltimore's associates, but could not withstand the criticism over the appointments from the press and Dingell. A new panel, formed after the Dingell hearing, included Dr. Joseph M. Davie, senior vice president for research and development at G. D. Searle in Skokie, Illinois; Dr. Hugh McDevitt, professor and chairman of the department of medical microbiology at Stanford University; and Dr. Ursula B. Storb, a professor in the department of molecular genetics and cell biology at the University of Chicago. Davie was named the chairman. Dr. Bruce A. Maurer, the executive secretary of the virology study sector of the NIH Division of Research Grants, served as the panel's executive secretary.

The investigation was further delayed by the need to use the OER staff to prepare for the Dingell and Weiss hearings in April of 1988. But the NIH staff began its Boston interviews in May, focusing on the roles of the various co-authors in conducting the experiments and preparing the paper, as well as on the responses of the universities to Margot O'Toole's challenge.

The newly constituted scientific panel traveled to Boston in June to meet with O'Toole and the *Cell* authors to discuss the science and review Imanishi-Kari's notebooks and data. The Davie panel, perhaps still envisioning its work in the collegial mode that scientists are accustomed to, did not tape its interviews, and contemporaneous notes could not be found later. Consequently, when new questions came up, no one could be sure just what documents the panel members had reviewed and whether they had seen what they needed or what Imanishi-Kari chose to show them.

Initial drafts were written, more questions raised, further interviews conducted, and more drafts written. In the midst of the investigation, Baltimore, Imanishi-Kari, Weaver, and Reis submitted a letter to *Cell* acknowledging "three instances of misstatement" that they said were "not material alterations and do not affect the conclusions of the paper, which remain appropriate and have been the basis of further studies." The letter, published in the November 18, 1988, issue, said that the Bet-1 reagent did not react only with the transgene allotype; in Table 2, the data represented wells of hybridomas, not isolated clones as stated in the paper; and the unpublished data showing that the non-IgM wells were "mostly gamma2B" actually showed that though the majority were gamma, they were not necessarily gamma2B, and came from other experiments, not those depicted in Table 2 as stated. But the letter did not go far enough, according to the outside panel's findings.

Davie, McDevitt, and Storb were highly critical of the coauthors, but they did not find fraud, nor did they use the phrase "scientific misconduct." Their report said there were "significant errors of misstatement and omission, as well as lapses in scientific judgment and interlaboratory communication. However, no evidence of fraud, conscious misrepresentation, or manipulation of data was found."

The panel said it had determined through O'Toole's own experience, the data in the seventeen pages, and experiments by various members of Imanishi-Kari's lab that the Bet-1 reagent would, indeed, react with the mouse's own gene as well as with the transgene. Although the panel found that there was a "relative specificity" for the transgene, it could not determine if the scoring for the transgene was appropriate because data on some of the radioactivity counts were not recorded. "The clustering of values of 'positive' samples in the low numbers, however, suggests the cutoff might not have been appropriate," the panel wrote. "The implication is that the estimate of transgene expression (as measured by mu^a allotype expression) may be inaccurate. This could be a significant error if a large proportion of these clones are producing an endogenous immunoglobulin heavy chain and low levels of the transgenic heavy chain, i.e. are double producers."

The data in the seventeen pages, as well as Margot O'Toole's experience with the reagent, raised questions about the actual sensitivity of the Bet-1 reagent, the panel said. "Dr. Imanishi-Kari's explanation of [radioactive] damage being responsible for the variability observed with this reagent is reasonable but unproven," the panel added.

While the panel accepted Imanishi-Kari's explanation that the so-called normal mouse in the seventeen pages was actually a mistyped engineered mouse and that another normal was used as a control for Table 2, the outside experts were still critical of the table. The panel noted that, contrary to published statements, the data was based on experiments on culture fluid ("supernatant") from the plastic wells containing hybridomas prior to cloning and subcloning, or diluting in a fashion to produce individual cells. Because particular wells probably contained multiple cells, the total number of transgenic clones screened was then much higher than the paper said, and the percentage of those positive for the idiotype was likely lower than reported in Table 2.

The error in not distinguishing between single cells and wells of more than one cell was "serious," according to Davie, the chairman of the panel. "This was an important technical issue," Davie would later explain to the Dingell subcommittee, "because it provides a quantitative basis for determining the difference between transgenic animals and the normal animals, and it was a significant part of the paper."

Other unpublished subcloning analyses given to the panel by Imanishi-Kari, however, supported Table 2's contention that a high proportion of hybridomas from the engineered mice are positive for the 17.2.25 idiotype "while only a small percentage of these idiotype positive clones express a high level of the mu allotype of the introduced transgene." These unpublished data convinced the panel that Table 2 was essentially correct, according to Sanford's McDevitt.

In a meeting with Imanishi-Kari after the NIH report came out, McDevitt recalled the panel's first session with her in Boston, when they asked about Table 2. Why was there such a high frequency of endogenous cells positive for the idiotype? Panel members said Imanishi-Kari was open about the fact that the table actually reflected data involving multiple cells.

But if the wells contained multiple cells "they could be mixtures of all sorts of things. And that invalidates the quantitative conclusion of the paper," McDevitt told Imanishi-Kari. "The quantitative conclusion is that this is a frequent event. What has been verified in other laboratories is that it occurs but not as a frequent event. . . . So, we essentially said in a variety of ways, well, that settles it, the paper is wrong."

Imanishi-Kari later disagreed with the panel's conclusion that the subcloning data were important. But McDevitt insisted that whether or not the unpublished subcloning was important to the paper, he and the other panel members believed it to be.

"We could be wrong about that, but that was the conclusion we were coming to. We then adjourned. You left. We went to a restaurant. We discussed it in rather a glum tone," McDevitt said to Imanishi-Kari. "I went back to my hotel. There was a message to call you, which I did not respond to.

"When we came in the next morning, you were waiting for us," McDevitt continued, "and you said, 'I got the impression when we left yesterday that you thought that the fact that the data in Table 2 was based on wells and not clones invalidated the paper. So let me show you some data of subclones which validates that it is a frequent event.' "

In their report, the panel's experts added that besides the problem with the multiple-cell wells, Table 2 was also inaccurate in its claims that the 119 clones were tested for isotypes other than mu, although it noted that isotyping had been done for the Table 3 clones, and that that isotyping partially agreed with the statements incorrectly relating to Table 2.

Several errors were found in Table 3, but the panel put them down to "carelessness in the presentation of the data."

In general, the panel found that the paper's "factual and clerical inaccuracies were inadvertent errors of understanding and communication between the three principal authors"—Imanishi-Kari, Baltimore, and Weaver.

The panel decided that the co-authors' earlier letter to *Cell* was not sufficient, and recommended that they both offer the journal the additional subcloning data, and note that the data published in Table 2 were not from clones and that the isotype tests were not done. The panel also wanted them to submit a brief report to the journal about the problems in the sensitivity of the Bet-1 and idiotype reagents.

Davie, McDevitt, and Storb were not relentlessly critical, however. They concluded that they were "impressed by the amount of work done in support of the studies published in the paper in *Cell*, by the completeness of the records, and by the abilities of both Drs. Imanishi-Kari and Weaver to find, accurately interpret, and present data on experiments that were performed as much as three or four years earlier." No such kind words for Margot O'Toole, the person who first pursued questions about the paper to the detriment of her career, were in the report.

The panel sent its report to NIH on November 10, 1988; NIH passed on copies to the co-authors and to Margot O'Toole, as well as to Dingell's oversight subcommittee. Neither the co-authors nor O'Toole were happy with the findings, but they could not change the minds of the panel members.

"Where the panel is critical, it has based its criticism mainly on the form of our presentation of the data. It is where the panel members would substitute their judgment for our own that we take exception," wrote Baltimore, Weaver, and Imanishi-Kari.

Despite the statement that Table 2 was based on wells that

might have contained multiple cells rather than on clones, the scientists argued that the table "was the best way to summarize a large amount of data in easily accessible form." They also disputed that there were factual inaccuracies in the table.

They disagreed about submitting the subclone data as a replacement for Table 2, contending that the data were less significant than the original data on the wells.

As for the problems in the sensitivity of the Bet-1 and anti-idiotype tests, they said those issues have been discussed in the scientific literature before and there was no need to repeat them. They did acknowledge, however, that they had gone too far in stating that Bet-1 was absolutely specific in its detection of the transgene.

Thus, the panel's criticisms of the paper came down to different interpretations, Baltimore, Weaver, and Imanishi-Kari said. But whatever the inadequacies of the review, the panel was not arguing interpretations: McDevitt, Davie, and Storb thought the Table 2 well data were useless and would have dismissed the paper had Imanishi-Kari not submitted the subcloning data.

Despite the panel's criticism of the *Cell* authors, O'Toole was unsatisfied. She blasted the draft as "a wholly inadequate scientific analysis of the facts submitted to the panel."

"The report draws important conclusions from experiments that Dr. Imanishi-Kari stated had not been done," she wrote to Dr. M. Janet Newburgh, who succeeded Mary Miers as the OER's scientific-misconduct officer. "In fact, the recommendations of the panel are based in large part on such questionable experiments. These include, but are not limited to, the subcloning analyses of Table 2 cells and idiotype analyses of 9 of 18 Table 3 hybridomas."

O'Toole argued that no such subcloning data for Table 2 had been produced for the earlier investigations.

"I state categorically that Dr. Imanishi-Kari told me in May 1986 that no subcloning analysis of Table 2 hybridomas had been done," she wrote. "I further state with certainty that no such data was examined by the previous investigators either in the meetings with Drs. Huber and Wortis or, several weeks later, at the meeting with Drs. Eisen, Baltimore, and Weaver.

"These data, not shown to anyone at the time of my original challenge, are the only evidence that allows the panel to conclude that there is a (somewhat) higher than normal expression of endogenous idiotype-positive immunoglobulin in the Table 2 transgenic hybridomas. This was the central finding of the paper and the one that I challenged."

O'Toole also attacked the report for treating the co-authors' inaccurate claim that isotyping was done as merely "a proofreading error" and a result of poor communications between the Baltimore and Imanishi-Kari labs. And she was astonished that the panel had used her anomalous Bet-1 results to back up Figure 1.

The panel considered O'Toole's complaints but said it had already examined those issues during the investigation. The report was reviewed by senior NIH staffers, who were concerned about O'Toole's charges that the experiments providing the supporting data had not been done. The NIH officials asked Tufts's ad hoc panel—Wortis, Huber, and Woodland—about the charges, and they responded that they did review original subcloning data that supported Table 2.

Imanishi-Kari was also questioned; she confirmed that she had given the subcloning data to Wortis, Huber, and Woodland, as well as to O'Toole. "At no time did I say to Dr. O'Toole that the subcloning analysis of wells in Table 2 was not performed," Imanishi-Kari said.

Dr. Katherine Bick, NIH's deputy director for extramural research, added in a summary memo accompanying the report that criticism of the responses of MIT and Tufts "appears misdirected" because there had not been any allegations of fraud or misconduct. "Thus, the decision of Dean Brown to ask Dr. Eisen to try to resolve the matter could be viewed as going beyond commonly understood requirements of academic integrity and stewardship of public funds," Bick said.

However, Bick did note in the memo, which was addressed to NIH's Wyngaarden, that "MIT's policy that an allegation of fraud must be made before any formal investigative process is invoked could be viewed as an unfortunate impediment to the timely resolution of the issues raised in this case."

Wyngaarden concurred in the panel's recommendations. On January 31, 1989, he notified the co-authors, O'Toole, and the universities of the findings. He directed Baltimore, Imanishi-Kari, and Weaver to submit the proposed corrections to him before sending them to *Cell* so that NIH could review them for completeness—an unusual slap at the dignity of a Nobel laureate. He also had strong words for the three scientists, disputing that they had given due weight to the questions raised about their paper. He was apparently unimpressed by the Eisen meeting with Baltimore, Weaver, Imanishi-Kari, and O'Toole.

"It appears," he wrote to them, "that even though the allegations have been known to you and the other co-authors of the *Cell* paper at least since the spring of 1986, the co-authors never met to consider seriously the allegations or to reexamine the data to determine whether there might be some basis for the allegation. Such an analysis on the part of the paper's co-authors, followed by appropriate action to correct such errors or oversights, may well have made a full investigation unnecessary."

Despite O'Toole's complaints, Wyngaarden thought that the panel's report and recommendations for corrections would go a long way toward resolving the controversy. In fact, he considered the case closed. But O'Toole continued to press the NIH to see the raw data for herself, and the case became anything but closed when NIH tried to comply with her request for the underlying data for Figure 1. This was the graph that purportedly showed that the genetically engineered mice had a large elevation of the 17.2.25 idiotype that could be attributed to the transgene, and that the mice's own mu^b allotype also had the 17.2.25 idiotype. NIH then learned about the Imanishi-Kari laboratory's unorthodox data-handling practices.

Imanishi-Kari told one of Wyngaarden's aides that some of the data generated by her postdoc Moema Reis for Figure 1 was not listed in the lab notebook, but instead was recorded directly on the figure. She also added that she had reorganized some of her lab notebooks, which NIH understood to mean that she had done so in preparation for the various inquiries. Although neither the late recording nor the reorganization of the lab files proved to Wyngaarden that there was any misconduct, he nevertheless reopened the investigation and ordered a detailed audit of all of the data that supported the *Cell* paper.

NIH officials knew from Dingell's subcommittee staffers that the Secret Service had been called in to examine some of the notebooks. The Secret Service findings would require the second NIH review to take a more forensic approach. Wyngaarden and the panel members would learn, too, of an explosive September 1986 letter from Baltimore to Eisen, a revelation that would raise the scandal to another level of intensity.

TEN

Preparing for Battle

■

The crowd of scientists, reporters, and congressional staffers in the Rayburn Building's House Energy and Commerce Committee hearing room was tense on May 4, 1989, and the nervous buzz was growing to a roar.

Scientists had come from around the country in a show of support for Baltimore. Among those present were fellow Nobel laureates Daniel Nathans, the Johns Hopkins geneticist who was honored for his efforts in gene mapping, and Marshall Nirenberg, the NIH biochemist honored for his work in cracking the genetic code; Cambridge colleagues Phillip Sharp, Robert Weinberg, and Gerald Fink; Norton Zinder of Rockefeller University; Eric Kandell of Columbia; and Harold Varmus of the University of California at San Francisco. Baltimore's students from his lab at the Whitehead Institute trekked down from Boston, carrying a good-luck card signed by 257 Whitehead people.

The Baltimore partisans had also reserved a seat up front for Mark Ptashne of Harvard, one of Baltimore's friends and early supporters. But by the time of the May hearing, Ptashne already had grave doubts about the *Cell* paper and had gone to Capitol Hill on his own to learn more.

Ptashne had not been impressed by what he had seen of Stewart and Feder at an earlier conference, and had called Baltimore to tell him that he was in the clear. He had also offered to write a piece for *Science* explaining and defending the science in the *Cell* paper. But as Ptashne delved deeper into the paper, he became unhappy with it. It looked sloppy. He thought the paper obviously had serious problems. Although he attended the Dingell hearing, Ptashne became upset when a news article listed him as one of the group who had come to lend Baltimore support.

Most of the others in the crowd at the hearing, however, had gathered, no doubt, in part because of Baltimore's earlier letter. But other events, and some efforts by Baltimore's friends, had fired up the scientists to go to Washington.

In January, there had been a meeting of congressional staffers and scientists at Cold Spring Harbor Laboratory on Long Island. The closed-door session was an attempt to bring the two sides together to discuss the problems of scientific misconduct and fraud. The Cold Spring Harbor research institute is headed by Nobel laureate Dr. James D. Watson, renowned for the codiscovery, with Francis Crick, of the molecular structure of DNA. Watson had reportedly urged Baltimore to be conciliatory toward Dingell, but given the nature of the two combatants, conciliation was highly unlikely.

■

The January conference had all the benevolent feeling of two unlikely alien cultures encountering each other for the first time. *Science* reported that the "exchanges were feisty and heated." "It was a hats-off, hair-down kind of meeting," Norton Zinder of Rockefeller University told the journal afterward. The Baltimore case was a hot topic. (Baltimore himself did not attend.) But what really upset the scientists was when two congressional staffers brought up the Holocaust.

The first to do so was Diana Zuckerman, an aide to Representative Ted Weiss, whose subcommittee had beaten Dingell the year before in the race to hold hearings on scientific misconduct. She raised the issue in the context of studies on how educated people could let something like the Holocaust happen. When one scientist asked how Zuckerman could compare scientific misconduct to the Holocaust, she carefully replied that she had not done so. But Walter Stewart, by now on temporary detail to the Dingell panel, also took up the issue, drawing on Edmund Burke's remark "The only thing necessary for the triumph of evil is for good men to do nothing." The lesson was clear to Stewart: Good and honest scientists must be willing to act against bad science. He did not equate the Holocaust with fraud and misconduct, either. But by some accounts he was more excited and looser than Zuckerman, and much more disliked by the scientists in the audience, many of whom were Jewish. They were insulted, furious, and ready to protest.

Rather than bring the scientists and congressional staffers to some new understanding, the meeting "made most people feel worse," said Zinder, who had known Stewart since his studies at Rockefeller and thought he and Feder were loose cannons. "The congressional aides felt we were being arrogant. The scientists felt the aides don't listen to them. All the stereotypes were exacerbated."

The scientific community was also riled up by an April 18 missive sent by Baltimore's friend and colleague Phillip Sharp, director of the MIT Center for Cancer Research.

Apologizing for the "Dear Colleague" impersonality of his letter—"Logistics and urgency dictate this form"—Sharp called on his fellow scientists for help in countering Dingell's subcommittee, saying that Baltimore and his collaborators had been "vilified for their 'fraudulent' work, even though there was no scientific basis for such a statement." He noted that more congressional hearings were planned and that the subcommittee had also subpoenaed the notebooks, notes, and correspondence of Frank Costantini, the Columbia University biologist whose only connection with the *Cell* paper was that he had created the mice used in the experiments. And if that wasn't scary enough, the panel also demanded that the American Cancer Society, which funded some of Baltimore's work, turn over its correspondence with the Nobel laureate and Imanishi-Kari dating back to 1982.

"It seems obvious that the congressional subcommittee has decided to continue to hassle David and the other authors and this has serious implications for all of us," Sharp wrote. Including a sample letter and talking points, he urged the scientists to write and complain to their lawmakers and the subcommittee, and to ask their local newspapers to publish the letters or op-ed pieces before May 1. "If this works, we will have gotten our message out to a large and influential segment of the population in a timely way," Sharp wrote. "The fight won't end there, but it's a good beginning."

The sample letter, which Sharp smartly directed the scientists not to copy in toto, said: "It is difficult to fathom the motives behind the subcommittee's current actions. But I believe that to continue what many of us perceive to be a vendetta against honest scientists will cost our society dearly. . . . When such matters are brought before congressional panels, there is a clear danger that calm scientific judgment will be overshadowed."

Battle was also joined by Dr. Bernard Davis, professor emeritus of bacterial physiology at Harvard Medical School, who published a column in *The Wall Street Journal* for March 8. The column, under the title of "Fraud vs. Error: The Dingelling of Science," accused Dingell of politicizing science and intimidating scientists. "The congressional inquiry has become a crusade, punitively pursuing one of the most distinguished and productive biomedical scientists, David Baltimore of the Massachusetts Institute of Technology," Davis wrote. He concluded: "Neither science nor society will benefit from a paralytic legislative crusade for an unattainable degree of purity."

And two days before the May 4 hearing, *The New York Times* published an op-ed piece by Dr. Robert E. Pollack, then dean of Columbia University's Columbia College and professor of biological sciences. Pollack warned that Dingell was threatening the scientific method: "As I understand the Congressman's case, it is that published science must be free of error, and that error itself indicates bad faith and fraudulent intent," Pollack wrote. "This is wrong. Published error is at the heart of any real science. We scientists love to do experiments that show our colleagues to be wrong and, if they are any good, they love to show us to be wrong in turn. By this adversarial process, science reveals the way nature actually works."

Pollack concluded his column: "I would welcome a congressional initiative to deal with fraud as such, but I fear that the way Dr. Baltimore is being treated means that witch-hunts are in the offing."

While the scientists were gathering their colleagues up for the confrontation between Dingell and Baltimore, Baltimore's legal team in Washington also was busy trying to ensure that some friendly faces would appear at the subcommittee hearing. Initially, the Akin, Gump lawyers, whose Democratic credentials were impeccable, were running into trouble making any points with the members of the subcommittee. Although Dingell did not demand loyalty as much on investigative matters as he did on legislative issues, it would nevertheless be a rare Democrat who would cross him on his own oversight panel. Some subcommittee Democrats agreed not to join the attack on Baltimore, but no one would go to bat for him. So the lawyers needed help in lobbying the Republican members, who might be more likely to give a kind word or ask a leading question during the hearing.

About a month or so before the hearing, they brought in Russell Smith. An associate at the D.C. office of the New York law firm Willkie Farr & Gallagher, Smith was also part of the Washington revolving-door community. He had once worked as a counsel to the Republican members of Dingell's committee and would have a better shot at them. The legal team put together an information packet, and Smith helped open doors for Baltimore with the GOP members of the subcommittee and their staff.

But the lawyers were not simply buttonholing members of Congress. The Akin, Gump team also sifted through documents in response to the subcommittee's demands, discussed how best to frame the issues, wrote congressional testimony, and prepared Baltimore for the hearing. Including Baltimore's Cambridge team—his personal lawyer, Normand Smith, and a part-time public-relations assistant, Al Kildow—there were often twenty people or so in the room when the Akin, Gump lawyers called a meeting at their Dupont Circle offices.

Baltimore's team would soon learn how successful they were at reaching the members of the subcommittee, and how well the Nobel laureate would do in a face-off against Big John. If anyone could stand up to Dingell on his own territory it would be Baltimore, whose seeming arrogance had already impressed the subcommittee staffers so much that they were referring to him as "Lord Baltimore."

Dingell, whose subcommittee had already been swamped with angry letters from scientists stirred up by Baltimore and his friends, had actually thought, at least briefly, that this hearing might not be necessary. If Baltimore was confronted privately with evidence of date-tampering, he would quickly move to resolve the situation somehow—perhaps he'd discreetly retract the paper, or at least step back from supporting Imanishi-Kari. Dingell thought he was giving Baltimore an out.

The subcommittee staff and Secret Service agents John Hargett and Larry Stewart met with Baltimore and various lawyers for several hours on April 25, 1989, and with Imanishi-Kari and her lawyer two days later. They also briefed NIH officials as well. Baltimore and Bruce Singal, Imanishi-Kari's lawyer, would later complain that they were not given any written report; they felt that the briefing was designed to terrorize them. They received some documentation over the weekend before the May 4 hearing, but Singal would persistently demand fuller, formal Secret Service reports on those and other findings over the next two years, contending that Imanishi-Kari could not properly defend herself without the raw Secret Service data. Dingell's aides say the scientists were shown the original data as well as the exhibits' blowups.

When the Secret Service evidence was laid out, Baltimore turned ashen, according to some of the people at the briefing. They feared he would become ill right there. (However, one of Baltimore's lawyers said that he didn't recall Baltimore appearing ill.) As for Baltimore, he said that if he looked shaken, it was because even though he still believed in the paper and Imanishi-Kari, he did not know how to respond to a confusing Secret Service briefing. By all accounts, Baltimore did not respond to questions at the meeting.

The subcommittee staffers did not have to wait long for Baltimore's next move. On Tuesday, two days before the hearing, Baltimore made the media rounds in Washington. In the afternoon, he met with *The Wall Street Journal* and then with *The New York Times*. He spent the evening with more than a hundred science writers at the meeting of the D.C. Science Writers Association. On Wednesday he had lunch with *Washington Post* reporters.

Baltimore went on the offensive at the on-the-record, fairly friendly give-and-take session with the science writers, laying out the basic arguments that he would give the congressional subcommittee the next day.

He began in a folksy way by comparing his new hobby of windsurfing to a career in science, both giving the satisfaction of achieving a difficult goal. But, Baltimore said, "The price one pays for windsurfing thrills are sore muscles, skinned knees, and a nose with no skin left on it because of the sun. The price extracted for science seems to have gone up recently. It used to be that one paid for science well by sacrificing time that might be spent in other pursuits or by not having time with family and friends. Today the cost, at least to me, seems to be measured in damaged reputations, lawyers' fees, and long hours learning the ways of Congress. I guess you all know those ways, but they are not exactly the ways of Cambridge, Massachusetts."

Recognizing that the reporters had already learned about the Secret Service findings, Baltimore laid out his arguments and called the briefing a "charade." "They have shown me nothing that causes me to have any doubts at all about the veracity of the *Cell* paper," Baltimore told the reporters.

He said that while the agents said some of the pages were out of order and some dates were changed, they did not indicate that any data had been altered. "In fact, the pages which concerned them contained none of the data that actually contributed to the paper in question," Baltimore said. "It was all, in a sense, ancillary data. . . . The staff said nothing about questions they may have in regards to the science, which makes it somewhat difficult to respond to. They seem to think that just because they have done this analysis we should be intimidated into believing that what they have found is meaningful."

Baltimore urged the writers to attend the May 4 subcommittee hearing and to stay through to the end to hear the whole story of the altered dates and page sequences. He clearly wanted to make sure that they heard his and Imanishi-Kari's responses to what he was sure would be inflammatory charges.

He gave them a rundown of the affair, noting that he believed that O'Toole's challenges within MIT and Tufts were "healthy and proper. In fact, I encourage all young scientists in my own laboratory to question, to attack, and they do so. They do so sometimes with more energy and enthusiasm than we want. But at this juncture what was needed in order to decide on the merits of her arguments was a new experiment with a new experimental approach. Not more discussion."

Baltimore also called for sympathy for Imanishi-Kari, a scientist whose style, he conceded, was "a bit chaotic," but who now had the forces of Congress and the Secret Service unfairly arrayed against her. "She has practically no ability to defend herself and her reputation. No record can be clearly shown that she had done something wrong which I have no reason to believe," he said. "She deserves my support and the support of all scientists, for any of us could be in her shoes."

He had no such sympathy for Walter Stewart, who with his boss Ned Feder had stirred up so much trouble with the study of the seventeen pages that O'Toole had found. Now on detail to the subcommittee, Walter Stewart had led the staff and Secret Service briefing. In Stewart's world, Baltimore charged, "the truly creative would be driven from science or from America."

But Baltimore had a much more delicate public-relations task to perform in the briefing. A few days earlier, reporters from *Science* and *The New York Times* had reported on a September 9, 1986, letter they had obtained, from Baltimore to MIT's Herman Eisen. Taken by itself, the letter appeared to be evidence of a cover-up at MIT; even if interpreted more charitably, it was an embarrassing and astonishing letter for such a leader of the scientific community to have written.

■

The letter, written on Whitehead Institute letterhead and marked confidential, was in response to a discussion Baltimore had had with Eisen and "after much thought about the situation brought on by my collaboration with Thereza Imanishi-Kari. . . . The evidence that the Bet-1 antibody doesn't do as described in the paper is clear. Thereza's statement to you that she knew it all the time is a remarkable admission of guilt."

Neither Baltimore nor his postdoc, Weaver, knew that there had been any problems with the reagent. "Why Thereza chose to use the data and to mislead both of us and those who read the paper is beyond me." But he said the meaningless data from the Bet-1 tests did not change the paper, because Weaver had sequenced enough of the genes to get a qualitative picture.

Baltimore argued against retracting the paper, however, because it would hurt Weaver even though he had nothing to do with Imanishi-Kari's work and Weaver's own work was solid.

"I think that a retraction would harm the innocent and raise doubts about quite solid work. I think we should, however, acknowledge to colleagues that the Bet-1 results are not reliable and I, for one, will be skeptical of Thereza's work in the future." A copy was sent to Weaver.

In an attempt to mitigate the impact of the letter's publication, Baltimore told the science writers that Imanishi-Kari's poor command of English was responsible for the misunderstanding. Eisen "told me about it and instead of doing what I should have done, which was to get in contact with Thereza Imanishi-Kari and talk to her about it, I went home and began to stew about it. And without thinking the problem through I fired off a letter to Dr.

Eisen," he said, according to a transcript of the session. "And I'm confident that if I had a day or two more to think about it I would have realized that we must write a letter to *Cell* if that information was correct."

The talk to the science writers annoyed the subcommittee staffers, but what stunned them was the co-authors' next end-run around them. Unbeknownst to the subcommittee until nine-thirty the night before the May 4 hearing, Imanishi-Kari, Baltimore, Weaver, and their lawyers asked for and got a meeting with the NIH review panel. They seemed to want to set the stage for the next day's hearing and make sure that the panel understood that they did not respond to questions at the Secret Service briefing because they had received no written documentation then. And they apparently wanted to find out what the panel knew about the upcoming congressional hearing.

NIH chief Wyngaarden would later tell John Dingell that agency officials and the panel members had discussed the propriety of holding the session and decided that it was all right to go ahead. "We have reopened the investigation. It will involve continued attempts to determine what is true and correct, and we might as well begin."

On the NIH side, panel members Davie, McDevitt, and Storb were at the unusual meeting, as were NIH's legal adviser Robert Lanman and the panel's executive secretary, Bruce Maurer. Chairman Davie called the subcommittee that night to inform the staff what had gone on.

Bruce Singal, Imanishi-Kari's lawyer, started the session with the Davie panel by saying that he understood that NIH had been told that his client had no explanation for the Secret Service findings and was uncooperative. "She was dying to answer questions because she felt that what they were saying was of no merit whatsoever," he said, according to a transcript of the session. "At my instruction as her lawyer, because of the fact that we'd just received the information and we hadn't received enough information on which to make an appropriate response, I instructed her not to answer, and I just want that clear."

"They did give us to believe that they'd shown you everything," McDevitt said.

"No," Imanishi-Kari said.

Normand Smith, Baltimore's lawyer, added that the briefings for Baltimore and Imanishi-Kari were held separately and said that different things were said to each one. "When we compared notes," Smith said, "we obviously didn't get exactly the same story, which gives us great pause for what we're going to hear tomorrow. I'm not at all sure that what we saw was all they want to present tomorrow."

McDevitt noted that the panel had been briefed from nine-thirty A.M. to about six P.M. by the Secret Service. "We looked at all of the details, we checked the registers, where they said it said register and all that sort of stuff, and we were shown this . . ."

Baltimore jumped in: "You actually saw the . . . analysis. We never were shown. We were just told this one is on top of that one."

Lanman tried to move the group from its complaints about the Dingell subcommittee staff. "I'd like to, in view of the limited time, focus more on us trying to get information, because we're not really here to assess the subcommittee's procedures," he said.

Davie and McDevitt started out with the supplemental subcloning data, which they said were stronger data than those that had actually been published. "And these are specifically many of the pages, I can't say all of them, but many of the pages are the ones that they have questions on," Davie said. "What's your response to that? . . . Why would you publish not your best data and so on?"

Imanishi-Kari didn't answer directly. "First of all, think I want to ask you what do you understand that we want to say in Table 2, and then I can answer this question," she said. "What do you think that was our intention with Table 2?"

Davie responded that the table was intended to show that the transgenic and normal mice have different frequences of idiotype-positive hybridomas. "Right, right," Imanishi-Kari said. "It's the frequency." But she contended that the additional subcloning data were not as good, although her explanation wandered off.

"But when did you do the original wells that were in Table 2?" McDevitt asked. Imanishi-Kari pointed out that Table 2 was done by Moema Reis, and then, consulting the lab notebook, said that it was May 8, 1985. She also said that the experiments on the normal mouse were done July 3.

"This is the data that was replaced for the normal data and the seventeen pages," McDevitt said.

"Right," Imanishi-Kari said. "And this was one of the original arguments of Margot O'Toole, that this fusion was not done at the same time as the transgenic fusion that is in Table 2. I mean, she is right; we didn't do at the same time."

The group then got mired down in a discussion of the subclones; no one was sure if they were all talking about the same cells or about the plastic wells containing the cells.

"Now let me clarify something that keeps popping up, okay, and it is an enormous misunderstanding," Imanishi-Kari said. "I in my lab as a practice—I know you told me I'm wrong—as a practice when I get hybridomas, real hybridomas, whether they are wells or clones, when I first clone them I call them clones. When I clone again the same that was done once, I call them subclones. So when I say that . . . And when you start calling your question here, subclones, I had to repeat what you know I say, but it keeps popping out like a stupid, how do you say, stupid little thing."

McDevitt, however, returned to the issue of whether the supplemental subcloning data were better than what the co-authors had originally published. Baltimore took a stab at clarifying Imanishi-Kari's earlier explanation and said that the Table 2 experiment was designed only to determine the frequency of the positive idiotype tests. The scientists did not care about cloning the wells for that table, he said; the frequency analysis did not depend on whether there was a single cell or more than one in a well, since it was a qualitative analysis.

"But how can you get a frequency analysis of wells of unknown numbers per well?" Storb asked. "My second question was that in our previous interviews my feeling was that you had thought that they were clones, as we call then now."

"No, no," Baltimore responded. "My terminology, my use of the terminology is probably worse than Thereza's because I don't ordinarily deal with this kind of data. . . . I knew they were wells."

The panel members, Imanishi-Kari, and, to a lesser extent, Baltimore, continued to argue this issue for a while; they were having trouble reaching an understanding over what data they were actually talking about. McDevitt finally said that the subcommittee staff was focused on whether better data were actually prepared after the *Cell* paper was published.

"The logic of the thinking, and Joe and Ursula can correct me if I've interpreted it wrongly, is that this data was written over much earlier data," McDevitt said. "It's better data. It strongly supports the thesis of the paper, but it wasn't in the paper; therefore it was fraudulent; it was fabricated."

But Baltimore wasn't accepting that. "I've heard that argument; I find that obtuse."

"Well, that's why we're asking what the answer is," McDevitt said.

Baltimore brought the argument back to the frequency analysis, which he said was best done on the wells, even though the co-authors had misidentified the wells as single-cell clones. Storb said she finally understood, explaining that in subcloning, many of the cells are damaged. But McDevitt was still having none of it.

"I do not like the argument because if there are a hundred clones per well and [you] do frequency calculations off of wells, okay, you're going to get a falsely high . . . And you look at it, say, there's one clone in a hundred that is making a lot of IgG . . . you score it as a IgG. That's going to give you a falsely high estimate of frequency," McDevitt said, with Storb agreeing.

Baltimore then wanted to make sure that despite their differences on the issue, the panel did not believe that there was fraud. "Your disagreement, however, is a disagreement with our judgment and about what we wanted to do, and let me tell you something," Baltimore said. "I'd probably go with your judgment rather than mine, but I was working with Thereza at the time and this is what we decided to do. . . . I might ask, is that deceptive?"

McDevitt assured him that he was not thinking deception. But he pointed out that when the panel first learned that wells were involved in Table 2 and not individual clones as stated, he and his colleagues "decided the whole thing should be thrown out the window, and it was Thereza coming back the next day and saying, 'Well, we've got subcloning that supports it,' that convinced us that maybe there was something to the thesis."

Chairman Davie also assured Baltimore, Imanishi-Kari, and Weaver that the panel thought it had an "honest disagreement" with them, and then brought them back to the subcommittee's concerns. McDevitt was still unhappy with the concept of doing a frequency analysis on wells with multiple cells, arguing that it could give a falsely high estimate. The wells had to be diluted and subcloned. "In my mind," he said, "you've got to clear them out and see what you get."

When Imanishi-Kari again tried to explain, Davie interrupted: "I think we, you know, we are hashing over things that I think we already understand each other's position. I think we disagree, but it's an honest disagreement."

Baltimore was relieved. "That's all I want to make sure, that it's an honest disagreement."

The next issue that the subcommittee was bothered about, Davie said, was when pages actually were written and why it appeared that data sheets were out of order. Thereza said she wasn't surprised that the Secret Service might have found pages written after the experiments: She frequently organized and reorganized her records as she did experiments and analyzed the data, and when she had time to do it.

How long was the lag time? she was asked.

"It could be in some instances one day; it could be in some instances two weeks; it could be months because I'm quite sure I did a lot of recording of the original data into those pads," she said. "From the dates of the experiments that is written on the page [being questioned] it's about three months. And I am not surprised about that."

She explained how she started with a spiral notebook and branched out into filing counter tapes in manila folders. And when she knew that the NIH was going to investigate the paper, she tried to put her loose papers into the proper chronological order.

"But the recording of this, I agree absolutely, they were done at different times, and I did record old data after I had recorded much newer data," she said. "It all depended on how busy I was."

Storb asked if her recording could have lagged even a year or more for some data after newer data. Dingell's aides thought Imanishi-Kari took this question as a prompting. She responded: "It could. I cannot remember. . . . I know it's bad. These are my practice. . . . I'm not an accountant. For me what's important is the data, and it's important for the sake of the cells to know when they were originated and what kind of analysis at the time of the experiment, not the time of recording."

Davie told her that even though he understood her explanation, "the alternate explanation is one that they [presumably the subcommittee staff] want made. And it's very difficult for you until you prove otherwise."

The group again started discussing how Imanishi-Kari ordered her pages, and she again had trouble answering. But when they started explaining the difficulty with the October date changes, she grew angry.

"Thereza," Baltimore said, "I think I'd better make this exact. You're not going to believe this. What they're saying is that you actually dated them, that you made up something that never existed, that you wrote the date on one day after a time that you originally cloned, and then you changed it to produce enough time in the records for the subcloning to actually have taken place."

"I think that's what they were implying," McDevitt agreed.

"I understand all the inference. . . ." Imanishi-Kari said. "The only thing that I can say is that I know which experiments I did; I know exactly how these experiments were done, and I know how much fucking shitwork I did for this. . . . And if I like to fake, I would do a better job than this."

Baltimore then gave an incredible defense against the charge that the supplemental subcloning data, which had not been published in the original *Cell* paper, were used to avoid charges of misrepresentation. "I don't believe it [fraud] happened, but let's say it happened. You can put anything you want in your notebooks, can't you, and nobody's ever said that you can't."

McDevitt tried to explain that the data were supplied to the congressional subcommittee.

"It was under subpoena," Baltimore noted. "She had no choice but to supply them whatever was in there."

In other words: Made-up results for federally subsidized experiments can't be fraud if they remain in the scientist's notebook, and the scientist can't be blamed if a congressional committee then demands to see the fraudulent results.

Toward the end of the four-hour meeting, Baltimore asked the NIH panel for a favor. "If there is anything else that you can think of that came up in the process of your conversation [with the subcommittee staff] even if you dismissed it as uninteresting or whatever, we would love to know about it. . . ." McDevitt noted there was some convincing evidence of pages with widely separated dates having been written on top of each other.

"Will Stewart and Feder be there as congressional investigators, as NIH scientists, as NIH fraudbusters, as what?" Baltimore asked.

Lanman started to answer: "That's a problem that . . ."

Baltimore interjected with a finality that concluded the meeting: "That's a significant problem."

ELEVEN

The Confrontation

■

Gaveling the hearing to order shortly after ten A.M. on May 4, 1989, John Dingell knew he would have to take control from the beginning or David Baltimore would command the agenda, the news media, and public sympathy. The Nobel laureate was proving to be as good a practitioner as he of the scorched-earth policy of public relations.

Opening statements by Dingell and other members of the panel set the adversarial tone of the hearing, and let Baltimore, Imanishi-Kari, the NIH panel, and others who were scheduled to testify know just what they were in for on that very long day. Dingell started off by lecturing the scientists about the purpose of congressional oversight, which is "to shed light on problems of public importance. A case in point is scientific misconduct." The panel had examined some cases of proven misconduct that had not been handled well by the scientific establishment, the congressman noted. "The apparent unwillingness on the part of the scientific community to deal promptly and effectively with allegations of misconduct is unfair to both the accuser and to the accused."

Dingell told those assembled that the focus was still the ability of research institutions and the NIH to police themselves, but it was clear to Baltimore's supporters in the hearing room that the congressional investigators believed that Imanishi-Kari was guilty of fudging the data and that Baltimore was covering up for her.

Dingell did not want to get into whether the results of the *Cell* research were correct: His panel had "far better things to do than police science." Nonetheless, he said, Congress annually authorized several billion dollars in research funds for NIH alone, and "it is the responsibility of this committee to assure that this money is properly spent and that research institutions, including the NIH, which receive these funds, behave properly."

Dingell let the scientists know that he blamed Baltimore and his co-authors for the fact that this hearing was necessary. He had hoped that they would resolve the panel's questions during the Secret Service briefings. He also questioned why they went to the NIH panel the day before to discuss the Secret Service findings, yet refused to answer the subcommittee staff's questions. "If the co-authors had answered the subcommittee's questions last week, today's hearings might not have been necessary," Dingell said. "It is with regret that we proceed in this particular way, because it appears that we were afforded neither cooperation nor choice in the matter."

The Michigan bulldog also wasn't going to let his pummeling in the press and by the letter-writers go unanswered. He attacked critics who contended that Congress was not able to understand or raise questions about science. And he put the scientists on notice that he would treat questionable science in Cambridge, Boston, New York, Chicago, Pittsburgh, or Stanford no differently from military demands for billions of dollars for Star Wars technology.

"However, no one questions the ability of the Congress to deal with these issues when Utah scientists demand $25 million for a cold fusion experiment, or the Air Force needs $70 billion for the stealth bomber program, or hundreds of billions of dollars are requested for Star Wars, or when enormous sums of money are requested for enormous particle-accelerator programs in budgets which we are now contemplating," he told the audience.

Dingell criticized the NIH panel for praising the cooperation of Baltimore and the others while remaining silent on Margot O'Toole, whose initial questions eventually prompted the panel to conclude that there were serious errors in the *Cell* paper. In fact, he noted, a draft of the NIH report had included a line applauding her efforts, but for some mysterious reason this acknowledgment had been excised. Dingell would later demand explanations from the NIH and its scientific panel members.

"We know that error is inevitable in the process of scientific research, and we do not propose in this subcommittee at this time or any other to review the results of scientific research for accuracy," he told the audience. "That is a job that we hope is done by the appropriate agencies and appropriate peer review. . . .

"The refusal to investigate concerns by fellow scientists and to address questions with regard to known errors appears to be antithetical to good science. It also appears to be a matter of question as to whether good science can function when one finds that there are reported investigations and experiments which did not take place."

Representative Norman Lent of New York, a member of the oversight subcommittee and the ranking Republican on the full committee, gingerly took the side of Baltimore and his colleagues in his opening statement. He wanted to defend the scientists without attacking the young woman whistle-blower and seeming to support scientific misconduct. Also, according to Capitol Hill tradition, it would be unseemly to come out swinging against another congressman. Lent, who worked with Dingell to limit Clean Air Act restrictions and product-liability suits, was one of the Republicans to whom lawyer-lobbyist Russell Smith had brought Baltimore to give his version of events. Lent referred to the op-ed piece that Columbia College's Dean Pollack had written for *The New York Times*, warning of the legislative threat to science.

"I strongly urge my colleagues to approach this issue with caution," Lent said. "None of us would want to be party to any proceeding which would inadvertently deter important biomedical research nor would we want to leave uncorrected any systemic problems in promptly and thoroughly airing scientific matters."

The Long Island Republican also defended the scientists' refusal to answer questions from the subcommittee staff. "I do not think we should place any particular significance on the failure of the co-authors to take advantage of the committee's offer that they respond," he said. "After all, that opportunity was only afforded after this hearing was scheduled and the adversarial approach had been adopted."

The subcommittee's ranking Republican, Thomas J. Bliley, Jr., also sounded a cautionary note. While agreeing that the panel had the obligation to ensure that the NIH and research institutions adequately dealt with scientific-misconduct charges, Bliley said that "we should not put such a burden on the scientific community or on individual scientists that important research work is discouraged."

Before Baltimore and Imanishi-Kari had their chance to testify before the subcommittee, the panel heard from the NIH and the Secret Service.

James Wyngaarden began his testimony with a nod to O'Toole and the subcommittee staff. "We are indebted to both Dr. O'Toole and the staff of this subcommittee for their interest and perseverance in seeking a complete resolution of this case," he said. "We look forward to their continuing cooperation as we pursue our investigation."

He told the panel that the Department of Health and Human Services, which oversees the NIH, was establishing regulations to make sure that research institutions dealt forthrightly with allegations of misconduct. He announced the formation of the NIH Office of Scientific Integrity and of the Office of Scientific Integrity Review within the HHS office of the assistant secretary for health. (Robert Charrow, the deputy general counsel of HHS who had been holding up the establishment of OSI, had finally left the government.)

"We have attempted to be responsive to the issues which the Congress as well as individuals and other organizations have brought to our attention," Wyngaarden said. He conceded, however, that "perhaps because of the very need for scientists to be able to trust each other and to build on the work of others, it has been difficult for much of the scientific community to accept the fact that incidents of scientific misconduct do occur, are a serious problem, and must be dealt with firmly."

Although members of Congress generally do their own interrogations with the help of staff-written lists of questions, it is not unusual for an aide to do some of the questioning, especially when particularly complicated or technical issues are involved. For this hearing, Dingell staffer Bruce Chafin, who was blessed with a phenomenal memory, led the questioning of the NIH scientific panel. In response to a Chafin query about how the panel thought the paper's error in referring to the items in Table 2 as clones had come about, Davie replied that it had been a communication problem, exacerbated by Imanishi-Kari's apparent language difficulty. "As has been pointed out, English is not her first language," Davie said. "Even yesterday, when we spent four hours with Dr. Imanishi-Kari, we really, at several times, had to discuss what she meant when she used particular words. It's a continuing problem that we have." Actually the only real problem the panel had in terms of understanding particular words was when Imanishi-Kari tried to explain her usage of the terms "clones," "subclones," and "wells."

But what about the paper's claim that isotyping had been done on the Table 2 hybridomas when in fact the experiment had not been done? "Is that misconduct?" Chafin pressed.

"I believe it is misconduct," Davie answered.

"Did you call it misconduct in your report?" Chafin asked.

"We called it a serious error," Davie replied. Davie tried to explain that it was difficult to characterize the mistake and that there was a subtle difference between misconduct and serious error. Nonetheless, he continued, "I'm not sure I can defend it at this juncture."

Representative Wyden jumped in and asked why, then, the failure to do the isotyping wasn't described as scientific misconduct in the panel's report.

"If you call misconduct an intent to deceive or to do wrong, that's a problem," Davie said. "That's clearly misconduct. It was not possible for us to be sure that they intended to deceive in that way. Clearly it was an error, but we are talking about a sentence, a single sentence in the paper, which says that this experiment was done on animals in Table 2, when, in fact, that experiment was done with different animals. . . . We are really talking not about something that was made out of whole cloth."

Davie finally reversed himself and said he did not believe there was any misconduct.

Concerned about the fate of whistle-blowers, Dingell elicited from the panel members and Wyngaarden statements that Margot O'Toole had acted appropriately in raising her questions about the paper. But, he said, "she's essentially been driven out. She has no future in research at this time, does she?"

"No, I can't tell you that either, but—" Wyngaarden tried to reply.

"Can't tell me that. How about you, Dr. Davie and Dr. Storb and your associates there on the panel?" Dingell said. "Can you tell us that she has a future?"

As scientists in the audience snickered, Storb said she thought O'Toole could have a future in science "if she so chooses." McDevitt added that he did not know if she had actually tried to get another job and had been turned down.

Dingell was not satisfied, however, and questioned Wyngaarden and the panel minutely on how statements that Margot O'Toole's behavior had been correct had been eliminated from the panel's report on the case. Dingell and his investigators seemed to know more about the internal workings of the NIH probe than the NIH scientists and consultants did.

"All I can say is that I did not excise them. I do not know," replied Wyngaarden.

Dingell turned to Davie. "Do you want to tell us about that?"

"Well, clearly," Davie said, "the report went through several versions. We all obviously wrote our own . . ." But Dingell was impatient: "Let us address my specific question, if you please."

Although Davie couldn't answer, Dingell continued to torture him and his colleagues with questions about the editing process and whether it was possible that NIH officials had removed the statements. No one could answer him.

"Well, nobody knows who changed it. They say that the NIH staff reviewed. They don't remember doing it. That means that it must have been NIH staff or some other person, perhaps somebody in the dark of night," Dingell said. "I'm just trying to find out how, and what is the integrity of these independent panels that you set up."

Following up Dingell's line of questioning, Wyden asked Davie whether the panel had known about Baltimore's September 1986 letter to Herman Eisen, in which he described Imanishi-Kari's Bet-1 statement to Eisen as "a remarkable admission of guilt."

Davie said the panel did not know of the letter during its deliberations, and, in fact, had learned of it only the week before, from the subcommittee staff. That would have been about the same time that *Science* and *The New York Times* revealed the letter's existence. Wyngaarden said he learned of it on April 19 or 20, 1989, at a meeting in his office.

The NIH got the letter in May 1988, when it and the subcommittee sought documents from the MIT and Tufts scientists, but the document never made its way to the NIH chief or the investigating panel. However, Wyngaarden said that was because the NIH staffer who knew of the letter was satisfied that it was no longer meaningful.

Wyden criticized the failure to alert Wyngaarden and the panel of the letter as a "very significant shortcoming."

But Wyngaarden said that while he initially thought the letter was "very damaging," he changed his mind later after he had had "a chance to review all the collateral material."

Lent led Wyngaarden through a series of questions designed to bolster the Davie report's contention that while serious errors were made, there was no fraud. With respect to Figure 1 and to the data written not in the lab books but directly on the figure, Lent asked, "You did not mean to infer that Dr. Imanishi-Kari made that up? It was there; it was not in the record book; it was unorthodoxically [*sic*] written . . . directly on the figure; is that correct?"

Wyngaarden agreed, noting that Imanishi-Kari's postdoc Moema Reis said that she did the experiment and wrote the data directly on the figure.

"You are satisfied that the experiment was done, even though you did not find it recorded in one of the thousand pages of notes?" Lent pressed.

"We are satisfied that we have an explanation for what happened," the NIH director replied.

Lent continued with friendly, easy questions about the scientific process of trial and error. He concluded, "And nothing we have talked about here today undermines the efficacy of this paper. The mistake was a non-material mistake and did not bear on the end result of the paper, is that correct?"

"That is the impression I have from the panel's report, yes sir," Wyngaarden said.

Finally, after three hours of questioning and lecturing, Dingell agreed to a rare lunch break. But after the hour recess, Baltimore and Imanishi-Kari still had to wait their turn while John Hargett, Larry Stewart, and Steve Herzog gave the Secret Service report on the date alterations and on the composite photograph that had been published. Before Imanishi-Kari could speak about the dates, Representative Wyden used her prepared testimony—which congressional committees routinely require witnesses to supply in advance—in questioning the Secret Service agents.

Wyden noted that, in response to the finding that five pages containing 1984 dates were actually written in 1986, Imanishi-Kari offered two possible explanations. One was that she had the 1984 data sheets out when she was writing down the 1986 data; in that scenario, the 1986 data sheet would have been placed on top of the

1984 page when she started recording the 1986 data. "Now, you've told us that not only do you have this indentation tying them together, it's my understanding that you have various ink pads and sequential analyses which, in your professional opinion, says that this is not a possibility. Is that correct?"

Hargett: "That's correct, sir."

Wyden: "So, is it, then, Mr. Hargett, your professional opinion that it could not have happened in the first manner in which Dr. Imanishi-Kari has described?"

Hargett: "It would not be plausible at all."

Imanishi-Kari's second suggestion, Wyden said, was that she actually recorded the data on a page purportedly from 1984 two years after that date. The agents agreed that was more likely.

Then Wyden turned to Dr. Davie and asked whether it was normal for scientists to write up lab records two years after the experiments had been conducted, to which Davie had to say no.

Although Wyden tried to question the panel about other changes in dates and their importance to the *Cell* paper, the scientists had trouble answering, noting that they, Baltimore, and Imanishi-Kari did not have a complete written Secret Service report on the findings.

"We tried to find out in as many specific examples as we could, exactly what had gone on," McDevitt answered. "Where we could remember it and ask, we got reasonably good explanations."

Hargett and Larry Stewart, who were sitting at the witness table with the NIH panel, were surprised by these answers; they thought the panelists had been persuaded by the Secret Service briefing that there were serious problems with the paper's dates. "Who are these people?" they whispered to each other.

Lent questioned the Secret Service agents about the availability of a written report. This, of course, was a point that Imanishi-Kari's attorney brought up many times. There was no formal written report, Hargett said, although there was an outline of some of the indentation analysis.

"We were never requested to do it, and we were also told that our testimony would come in the way of an oral address to the subcommittee," Hargett said.

In response to Lent's questions, Hargett noted that he did not know anything about the science of the *Cell* paper. But he added,

"I certainly have reason to question the preparation of these pages. Whether or not the science justifies the corrections that have occurred is beyond me."

Steve Herzog also said that he was not implying that Weaver's composite photograph of the genes, Figure 4, was intended to deceive, only that "something has been removed, for whatever reason."

That "something," Davie testified, was a representation of a gene whose size was consistent with that of the transgene. The display was intended to show that the individual hybridomas were expressing the idiotype of the transgene without producing the rest of the transgene. "It had relevance to the conclusion, but in that same figure, there were at least two other cells that did not have any hint that there was that kind of possible second product in those particular cells," Davie said. "Whether that individual band was there or not would not have changed the overall conclusion of the study."

Davie added that the negative stripping that occurred is common and accepted practice in preparing photos for scientific papers. But he also said that the panel members would probably have thought it better to include in the paper a notation explaining what had been done. In fact, at the meeting with the co-authors the day before, McDevitt said that the panel members had told the Secret Service that composites are common but that it is usual to note when there are different exposure times.

Dingell, however, wanted to put a finer point on the issue. "What you have here is a picture that conveys different information than the original picture; is that not so?"

Davie: "Yes."

Dingell: "Was that misleading?"

Davie: "In a sense, yes."

But Dingell was having none of that ambiguity. "Well, it either is misleading or it's not misleading."

Davie: "There are different reasons for choosing a particular—"

Dingell: "No, no, no. It is either misleading, or it is not misleading. Is it misleading or is it not misleading?"

Davie: "It's misleading." Then Davie tried to explain that a scientist chooses the exposure that "gives the information that you're attempting to transmit."

This gave Dingell the opening for one of his lectures. "It's a funny thing," he said, "when I was back in college, I was a chemist. I was always taught that we were to present the data the way we got it; not to doctor it; not to change it. We were taught that it was good science to present this data the way we got it. Was I taught wrong?"

But while Davie was forced to agree with Dingell, he, nonetheless, pressed on. "This is a picture of radioactivity. There is no right exposure or wrong exposure." Davie conceded, however, that the scientists should have tested the "smudge" further to determine whether it was a transgene or an endogenous gene.

Finally, Dingell moved on to Baltimore, Imanishi-Kari, and Weaver. Baltimore was the first of the co-authors to testify. Although he later said that he was not deliberately seeking a confrontation, Baltimore came on strong from the beginning.

"You in your opening statement indicated that you thought this was a matter that should be discussed among us. Well, I agree entirely and this is not the forum for it," Baltimore declared.

O'Toole, he said, had acted properly by raising her questions at MIT and Tufts. He believed that new experiments needed to be done to further explore her arguments. But Walter Stewart and Ned Feder, Baltimore contended, had waged a vitriolic crusade against him and then convinced the subcommittee staff to join them in "this relentless campaign."

"At a public conference, while in the service of this subcommittee, Stewart made the loathesome comparison of scientific fraud of which he accuses me, to the Nazi Holocaust," Baltimore said. "On Tuesday of last week, we learned that even the U.S. Secret Service had been brought in to bolster their allegations."

But that Secret Service briefing "was designed to terrify without providing any substance," Baltimore said. He presented an affidavit from Benjamin Lewin, the editor of *Cell*, affirming that the use of composite photos such as the one in the paper is a customary practice that doesn't require a footnote.

"I must tell you, Mr. Chairman, I am very troubled about how this situation got so out of hand. I have a very real concern that American science can easily become the victim of this kind of government inquiry," Baltimore said. "Professor Imanishi-Kari is also a victim. . . . She has difficulty communicating in English, as the history of this controversy painfully shows. She deserves my support and the support of all scientists, for any of them could be in her shoes."

Baltimore then explained the September 1986 letter to Eisen, noting that he was not proud of it. "This was bad judgment, and I'm confident that if I had thought about it a little longer, I would have wanted to write a letter to *Cell*," Baltimore said. "I trust that the subcommittee understands my profound regret for writing this letter, and will accept the statement that I fully understand that when a serious error has been made, it must be fully acknowledged."

Baltimore categorically denied that the scientists had committed fraud or misrepresented data. And the science of the *Cell* paper? "No result of the paper has been proven wrong. A number of results have been replicated. Significant progress has been achieved by scientists building on the paper's conclusions." For the record, Baltimore supplied a list of five studies that he said built on the paper's conclusions.

Unsurprisingly, this carefully stated answer, that of Henry Wortis the following week, and statements by other supporters led some members of Congress as well as the scientific community to believe that the paper's central conclusion on idiotypic mimicry had been confirmed. That would be a powerful argument against suggestions that part of the science had been made up.

But a few scientists, like Harvard's Mark Ptashne, read those other papers and determined that while they might have confirmed aspects of the *Cell* experiments, they did not replicate the paper's central conclusion.

In his written statement submitted to the subcommittee, Baltimore also discussed his view of collaboration, an opinion remarkable for a scientist trained to be skeptical:

"Given Professor Imanishi-Kari's status as an independent investigator, I did not question closely her data, and I saw only occasional samples," Baltimore wrote. "It would have been inappropriate for me to have quizzed her more closely as it would have implied a lack of trust and belief in her. A collaborative effort of this sort requires a high degree of mutual trust or collaboration will break down and no benefit will come from the work."

Following Baltimore was his former postdoc, David Weaver. Although Weaver was the paper's lead author, he had deferred to Baltimore in discussing it once the controversy broke. Now he spoke briefly, explaining that the paper was based on two complementary approaches: Imanishi-Kari's lab did the serum work, detecting antibody proteins, while Weaver, in Baltimore's lab, did the molecular biology, looking at the RNA and DNA of the cells.

"It is important to reemphasize that the redundancy of these two different approaches was a cross-check on the science," Weaver said. "It made the publication stronger, and it is the sort of activity that should be actively encouraged and not discouraged."

Weaver said that while he had discussed his work with O'Toole before the paper's publication, she had not brought any of her concerns to his attention. He went on to say that it was his impression that her questions had been resolved by the Eisen meeting.

Weaver also argued that the Secret Service tests of the X-ray film for Figure 4 actually massively overexposed the film and masked the true size of the gene. "The NIH panel examined all of the X rays used for Figure 4 and raised no issue with regard to this methodology. I think that an emphasis on this point demonstrates a lack of perspective and understanding about scientific data," Weaver said. "The basic fact remains that the data for Figure 4 strongly supports our conclusion."

In Weaver's prepared statement put in the subcommittee's record, he added that Baltimore supervised his work, reviewed his data on a regular basis, and shared in the analysis of the results. And, he noted, the data for the paper were routinely communicated to other scientists through seminars, meetings, and conversations before the paper was actually drafted and published.

Imanishi-Kari's opening comments were even briefer than Weaver's. She apologized in advance for any statement that might confuse the lawmakers. She said that she was shocked to find out from Henry Wortis of Tufts that Margot O'Toole had concerns about the paper, and she was shocked by the Secret Service briefing. "It sounded to me like they were saying I made things up. My lawyer counseled me not to answer any questions there because it was all new and we should have a chance to understand the evidence and be able to respond to questions accurately. Now I have a chance to see some of these new charges and to me they make no sense.

"What they seem to be saying is that I am not a neat person," Imanishi-Kari said. "Well, that is true. I do keep my notes in what seems to others as a messy condition. But I know my notes. I know where they are and how to read them, and that's what's important."

But while her portrayal of herself as a perplexed scientist with messy record-keeping habits and difficulty in speaking English helped to counter the negative characterizations of her, Imanishi-Kari's most effective testimony gained much more public sympathy for her in the popular media. For personal reasons, she told Dingell, it was absolutely vital that her data and the paper be strictly accurate.

"These experiments are the ones that hopefully will guide scientists in trying to cure diseases of the immune system. One of these diseases is lupus, which is an autoimmune disease," she explained. "The data in the *Cell* paper and the related experiments in my lab can lead directly to a cure for this potentially fatal disease. . . . Mr. Chairman, I have lupus. My sister died from lupus. That was in my mind all the time I was doing my research. I have hoped all along that I could help provide insight that might lead to a cure for this disease. If I had fabricated data, it would have misled scientists, wasted their precious resources, and retarded their efforts to cure the disease that killed my sister and threatens me."

■

Fortunately for Dingell, the House bells rang right after Imanishi-Kari's revelation; a vote was being called. Dingell announced a brief recess to allow the lawmakers to go to the House floor and cast their votes. (Failure to vote usually gets thrown back at legislators at reelection time.) The recess would give some distance from Imanishi-Kari's painfully personal testimony, so subcommittee members would not appear to be attacking a very vulnerable woman.

When the panel returned, Representative Wyden avoided Imanishi-Kari, instead questioning Baltimore about the incriminating letter to Eisen, in which he is trying to protect David Weaver. Wyden noted Baltimore's explanation and apology, but continued, "My sense is, if this is what's done by our very finest—those who win the Nobel Prize—what does that say about scientific procedure?" Wyden said.

"Mr. Wyden, I can only continue to apologize," Baltimore responded. "It does not to me now represent good judgment, what I said in that letter, and I fully believe that when I had calmed down from a sense of outrage and loss of trust which was driving my feelings at that time, that I would have quickly seen that that was not an appropriate solution, and that I would have insisted, just as you're insisting, that if an error like that is found, it needs to be acknowledged, and as quickly as possible."

But what was Baltimore's responsibility as a noted scientist signing on to a paper?

That would be more difficult to answer, he said. When a paper is based on experiments solely from his lab, Baltimore said he takes "as full responsibility as I can for everything that occurs." But this was a collaborative enterprise, because he did not have the expertise to do the entire set of experiments. That was where Imanishi-Kari and her lab came in: "Not having that expertise, I also could not sensibly judge her work," he said. "I don't have enough experience to say that if Bet-1 doesn't work occasionally, that you don't use it or you do use it."

Baltimore added that because of his lack of expertise he could not second-guess her, even now, on how to set up the experiments.

He was unaware at the time of her rather haphazard data-recording practices. "I tended to meet her in her office, rather than her laboratory, and talk about scientific issues, and see data." But he launched a strong defense on her behalf: "It is certainly not a form of data handling that I would recommend to the scientific community in general. On the other hand, I know full well from my interactions with her, that with her understanding of the experimental system, which is quite remarkable, that she could reconstruct the history of her work in great detail and with the kinds of cross-checks in it that indicate that she's not misremembering things; that she has it all under control. . . .

"The Secret Service analysis, as elegant as it is, it seems to me simply proves that different scientists do their science in different ways," he continued. "I guess I encourage that. I encourage scientists to be individuals, and I'm less worried about somebody who is messy than I am about somebody who is not smart."

But why, Wyden asked, were the various investigators not told that the 1984 data weren't put together until 1986?

"I don't think that's a question I can answer," Baltimore replied, not mentioning that he knew his lawyer, Normand Smith, suggested to Imanishi-Kari that she put her papers in order before turning them over to NIH and the subcommittee.

Wyden also asked whether Baltimore was concerned about O'Toole's career prospects. His response was fairly cold: "I don't know much about Dr. O'Toole's scientific interests. I don't know what she wants to do with her life. I feel, as other people said, that in an appropriate laboratory setting, there's every reason why she should be able to continue her career. Since I've never had any responsibility, and basically don't know Margot O'Toole at all, I can't say any more about it at this point."

Wyden was not happy with what he heard from Baltimore or the NIH investigating panel. Although he said he was trying to avoid passing judgment on the science and he was "not interested in setting up a federal science police squad," Wyden said something needs to be done to "strengthen the scientific process."

His Republican colleague Alex McMillan, of North Carolina, mildly rebuked Wyden's statement that the paper's science was not at issue "because I think that's been implicit in the whole discussion." He defended the *Cell* paper and the authors, saying that the work "was done with integrity and possesses integrity."

Dingell went next, pressing Baltimore about his letter to Eisen, which he said indicated that Baltimore's instinct was not to go public. He quoted from the letter and asked the scientist why there was no letter or memo in his files negating the first one, and how he knew now that within two days the issue had been clarified.

"Because I remember the incident," Baltimore said, exasperated. "But I don't . . . prepare my files in a way that I expect other people to go through them, and so I don't generally send memos to the file. The file seems to be big enough without that."

And why couldn't Eisen produce anything in writing to indicate that the matter had been cleared up?

"It's because we use the telephone. We don't keep telephone logs, and we don't keep records of conversations," Baltimore said. "We're all friends and just, you know, kind of talk together. It's not a—we're not developing a paper trail."

Imanishi-Kari had been left alone for a while, but it was now her turn to be grilled by Dingell. He asked about the Secret Service findings: records dated 1984 for experiments done in 1985; a page dated 1985 done before a page dated 1984. "I say, can you explain to us how this occurred, how handwritten data was transcribed from . . . an experiment which occurred more than six months previous?" Dingell asked.

"I cannot give you an accurate answer to that because I don't remember," Imanishi-Kari said. "It was a practice either to transcribe data from the printouts or it was a practice to record. Sometimes I did one. Sometimes, another. Some people in my lab like Margot O'Toole never cut and paste tape. She always transcribed.

"I am not a systematic person. Sometimes I do, sometimes I don't."

Dingell seemed wearied, not only by the late hour—it was after six P.M.—but by this response. He interrupted and again criticized her, Baltimore, and Weaver for not responding to the subcommittee staff and the Secret Service at the briefing the week before and for not adequately answering the panel's questions. He argued that, despite their contentions to the contrary, the scientists had been given all the Secret Service information that was presented to the subcommittee.

"Your response, Dr. Baltimore," Dingell said, "was a rather ringing attack upon this committee and essentially an allegation that you had been charged with fraud, which happens to be untrue, and the fact or rather added allegation which indicated that some of the persons involved in these matters were behaving in a fashion worthy of Hitler and the Holocaust.

"Now I will tell you that I take umbrage, first of all, at those statements," Dingell said.

"Mr. Dingell, might I respond?" Baltimore tried to interrupt. But Dingell wasn't finished.

"Your response leaves us with about the same questions which we had when we commenced the hearings and possibly a couple more," Big John said. "I am not satisfied with any of the statements or any of the information which has been received. The committee is going to pursue these matters further and we thank you all for your assistance to us—"

"Mr. Dingell," Baltimore called out again, unafraid to confront the congressman. It was his reputation that had been attacked in the *Globe* article. And even though he had not been at the Cold Spring Harbor meeting, as a Jew he was outraged by the reference to the Holocaust. "Might I respond? Please?"

He had indeed been charged with fraud by Dingell's aide Peter Stockton, Baltimore insisted, waving a copy of the *Boston Globe* article. Then he quoted an article from *Science* about the Cold Spring Harbor meeting, which said that Walter Stewart wrote "Holocaust" on the blackboard. Baltimore read aloud: " 'By this, Stewart meant that the problems of scientific cheating were being ignored by many researchers, who like some Germans, dealt with the problem by looking the other way.' . . . So my charges, sir, are not made lightly."

Baltimore continued that, indeed, the co-authors had not seen all of the Secret Service data, including the cluster analysis of which sheets of paper were closer together. "We are people who think about statistics. We think about how do you validate scientific data," Baltimore said. "It's as if we presented you with the *Cell* paper and asked you to analyze it on the spot. That would be pretty unfair to do and that's the position we were in."

Dingell tried to point out that he had also sent Baltimore a letter informing him of the purpose of the hearing and asking for any additional information that might be useful. But Baltimore said he did not know what Dingell was talking about, and Dingell decided to cut the discussion off. It was pretty obvious that he had lost control of the hearing and that Baltimore was winning some big points.

"In any event, it is not the intention of the Chair to quibble further," Dingell said. "The Chair is simply going to observe that serious questions have been raised here today about scientific research. And the committee is going to pursue these matters to make sure that where questions arise, there is a proper mechanism for addressing those matters internally inside the scientific community."

It was six-forty P.M., eight hours after the hearing had begun, when he adjourned the panel. The scientists in the audience jumped up to hug and kiss Baltimore and Imanishi-Kari. They felt that their colleagues had gone head to head with Big John and had won. Baltimore was overwhelmed. "It was heartwarming to get such support," he said when things had quieted down. "I needed it."

TWELVE

The Third Hearing

■

Although reporters covered the Baltimore hearing and noted Imanishi-Kari's heart-rending statement about her sister's death and her own case of lupus, the news media generally held their fire until after the second subcommittee hearing five days later. On the docket were O'Toole, again; Wortis and Huber of Tufts; Eisen of MIT; and various university officials. Although the atmosphere was a little less charged than it had been at Baltimore's confrontation, the lines were clearly drawn from the beginning.

Dingell opened the hearing by criticizing the NIH: Its investigatory panel, he said, wasn't able to determine if there was fraud because it had not tried to ascertain the intent of the *Cell* authors. He noted that the NIH did not include in its report or in letters to the universities any acknowledgment of O'Toole's assistance in revealing serious errors in the paper.

Lent, however, used his opening remarks to warn against congressional interference in the issue. "If we allow leading-edge scientific work to suffer because of attacks and public efforts to discredit our best and brightest instead of encouraging sound research and informed debate by scientists, we will soon find ourselves with no scientists and nothing to debate or criticize," Lent said. "Criticism of existing academic institutions and their structures for resolving debates walks a thin line between legitimate oversight and attack."

O'Toole came in for harsh questioning by the Republicans on the subcommittee. Nonetheless, she spoke more confidently than she had in 1988, and was more willing to publicly accept her role as a whistle-blower. Just as Baltimore said he was standing up for all scientists who might be unfairly attacked by politicians, O'Toole's stance was that of a representative of scientists who try to question bad science. And she was taking her responsibility seriously.

"The scientists tell me that I have to show that I am right scientifically before they will accept outside criticism of the process I have gone through," she said. "I say to them, my knowledge of the facts of this case makes me absolutely confident that I will eventually prove my points beyond all shadow of a doubt. . . . The scientific community will then be left with a record of conduct that will not make it proud. . . .

"I stand for all those who raise questions about research practices. My battle now is to make sure those who follow me will get a fair and thorough hearing, that they will be allowed to examine the evidence and to see the reports."

O'Toole described again the process she had gone through. And she noted the snickering in the audience the week before when the subcommittee asked about her job possibilities.

"When was the last time they hired someone who eminent scientists describe as one who purloins data and lies under oath?" she said defiantly. "Who among them has ever hired anybody who reputable people say selectively copied notes to make trouble or support a particular point of view? Who would want in a laboratory someone who cannot accept an alternative interpretation of data, who cannot distinguish important issues from trivial issues, does not accept the reviews of eminent and qualified reviewers and who has unfairly besmirched the public's view of research? All these things are being said about me, and all these things are false."

O'Toole thanked the subcommittee members for their concern for her in the face of public criticism. Then she told them that she had informed the NIH scientific panel that the supplemental subcloning data it had relied on to bolster the inadequate Table 2 had not existed at the time of her challenge in 1986: She had asked Imanishi-Kari if the Table 2 wells had been subcloned and had been told no.

She pointed out that she told the NIH panel in 1988 that a data page now dated November 1984 had been shown to her on May 23, 1986, during the Tufts review. She said that this was the page, with her notations from the May meeting, that included data that Imanishi-Kari said she had just generated.

"The [NIH] panel paid no attention when I told them this," she said. "I did not know and neither did the panel that the Secret Service would later date this page to May 1986, contradicting the written day of May 1984.

"This episode demonstrates a beautiful fact of nature and a basic tenet of science: The only version of events that can fit all the evidence is the true version."

Dingell was a friendly questioner, prodding O'Toole to discuss the difficulties she faced after challenging the paper and what she thought of the Tufts and MIT reviews. "Do you believe Tufts and MIT adequately addressed your concerns?" he asked.

"No. I wanted a correction of the false statements," O'Toole replied. "They knew of the false statements and they refused to correct them."

Lent's questions were not so easy, even though he assured her that he did not doubt her good faith. He grilled her about statements by the other scientists that implied that the paper's central conclusion had been duplicated or that the paper's errors were of minor consequence. But O'Toole stood her ground.

"There were always findings in the *Cell* paper that I didn't challenge," she said, "and some of them have been confirmed. That is correct. The part of it that I challenged has not been confirmed."

Lent: "You agree as well, do you not, that the errors that may have been committed by Dr. Imanishi-Kari were not material to the thrust of the *Cell* paper, to the science of the *Cell* paper."

O'Toole: "They were material to the thrust of the *Cell* paper."

Lent: "You are saying the Stanford and Harvard and Brandeis [letters], NIH, MIT, and the Tufts examinations, which concluded otherwise, essentially that the paper was correct, are wrong?"

O'Toole: "The scientific papers do not address the issues I addressed scientifically."

For the record, Lent included letters to Baltimore from Leonore Herzenberg of Stanford, Philip Leder of Harvard, and Erik Selsing of Brandeis. What Lent did not note, but O'Toole did, was that the experiments done by those scientists did not confirm the central conclusion on idiotypic mimicry. In fact, Selsing, who had since moved to Tufts and was working in Wortis's department, stated, "In contrast to one of your hypothesis [*sic*], we do not believe that our work provides evidence for idiotypic mimicry." And Leonore Herzenberg, who praised Baltimore's work, thought O'Toole's criticisms of the paper's science were valid, even though the Stanford scientist believed honest error, not fraud, was the problem.

Representative Doug Walgren, a Democrat from Pennsylvania, asked for O'Toole's suggestions for resolving the problem of defensiveness within small scientific communities. Her answer was more openness, more collegial arguments over the meaning of particular experimental results. She noted that Baltimore had testified the week before that he had not examined the raw data because that would have shown a lack of trust in his colleague. "That is where Dr. Baltimore and I differ," O'Toole said. "He did not even examine it after I told him there was something wrong with it. The reason is that scientists feel they cannot challenge each other without calling trust into question.

"My reply is you become a scientist because you have decided not to take things on trust. You have decided you are going to examine the facts yourself and make your own conclusions. In a collaborative study, especially, people should be involved in the steps of taking the data and seeing how to present it."

If O'Toole had not shocked the scientific community before, she was about to really rattle it. Scientists, she said, should not consider their notebooks for taxpayer-funded work to be personal diaries. Her response, too, reflected a growth in her own thinking and how far she had traveled from being a typical team player:

"I think they should be left open, and we should go over raw data with each other more often, and this should be more accepted among us," she said. "My biggest battle was, believe it or not, people were absolutely shocked that I had looked at the data. . . . The notebooks were the science. The notebooks were what I was basing my own work on and I didn't see anything wrong with looking at them. I didn't think about it until they fell open in front of me, but I now think it would be my duty to examine the data for the work that forms the basis of something I was doing every single day, my own work. It is like building on an unsound foundation if you don't."

In response to a question from Dingell about how the process of getting legitimate challenges heard can be changed, O'Toole again called for more accountability and openness. "People say it is just one paper, and I say to them, if you can't even get one paper corrected, what does that say about the whole system?"

Alex McMillan seemed, like Norman Lent, to be trying to shake O'Toole's story. He pressed her to say that the core of her disagreement with Imanishi-Kari was her own bad results with the Bet-1 reagent, but Margot was firm.

"No, sir, that's not true. I found fault with the paper because when I was looking for some mouse breeding records, I found the records for one of the published experiments. That's why I found fault with the paper. . . . I never would assume my results were right and another person's results were wrong, and I didn't challenge them on that basis. . . . I looked at an experiment they had done, they had published, and I said it had been presented misleadingly."

"I'm a little confused," McMillan said. "You said they didn't have adequate data to back up their conclusions."

Exactly, O'Toole replied. "The raw data did not in important ways correspond to the published data and what they said about the raw data. The NIH panel looked into that, said I was right, but found other data that did support those conclusions. And what I say now is these other data were stated not to have existed at the time of my challenge."

But why did she not publish her challenge? McMillan asked.

O'Toole said she did not try to do so because Baltimore had told her that he would counterpublish and say she was wrong. "It didn't seem that it would do any good, because I thought he would be believed and I would not be, and I think that point has been proved pretty conclusively."

McMillan wasn't finished, however. He criticized her for not publishing and then "challenging the modus operandi of those who published the paper. It seems more personal than on the point of the scientific results."

O'Toole was incredulous. "I beg your pardon? I went to the authors with their own data and I said look, there are these things wrong with these data—you don't think that was the right thing to do?"

She pointed out that Stewart and Feder had indeed written a manuscript criticizing the *Cell* paper and could not get the authors to review the data with them, nor could they get their criticism published.

O'Toole did agree with McMillan that Capitol Hill was not the appropriate forum to discuss the facts and issues raised by the *Cell* paper. However, she said, "it's just that we are stuck here because we couldn't work out a better one."

■

The remainder of the hearing was devoted to the Tufts and MIT scientists who reviewed O'Toole's questions, and to university officials.

First up was Martin Flax, chairman of the Tufts department of pathology and pathologist-in-chief at the New England Medical Center. He had known O'Toole for more than ten years (she had earned her doctorate at Tufts). He said Tufts had recruited Imanishi-Kari in the summer of 1985, to move over July 1, 1986. In December 1985, he sent her a letter confirming Tufts's intent to hire her.

Although, Flax said, he initially told O'Toole to take her complaints to MIT, where the research had been done, he later encouraged her to proceed with the Wortis review. At the same time, he asked the late Sidney Leskowitz, the university's senior immunologist, to reexamine Imanishi-Kari's record with her supervisers, colleagues, and students. Eisen and Baltimore were among those interviewed. The general view, Flax said, was that Imanishi-Kari was a first-rate scientist who was moderately productive and was a helpful colleague. The pathology department's immunology group was convened; it unanimously voted to recommend her appointment.

"I can assure this committee that Dr. O'Toole's disagreements with Dr. Imanishi-Kari were not taken lightly. . . . It was certainly not in our interest to bring on board as a colleague someone who falsified or misrepresented data or was secretive and distrusted by her peers, colleagues, and students," Flax said. In fact, he noted, Imanishi-Kari moved into the lab adjacent to his own office and he saw her frequently discussing experiments openly with students and colleagues.

And, Flax added, Peter Brodeur, O'Toole's husband, remained "a very valued member" of the pathology department.

Brigette Huber, who had been on O'Toole's thesis committee, implied in her testimony that Margot was the one who was secretive. O'Toole's allegations, she said, were "based on informa-

Dr. Margot O'Toole (Ruth Fremson/*Washington Times*)

■

Dr. Thereza Imanishi-Kari (© 1993 Seth Resnick)

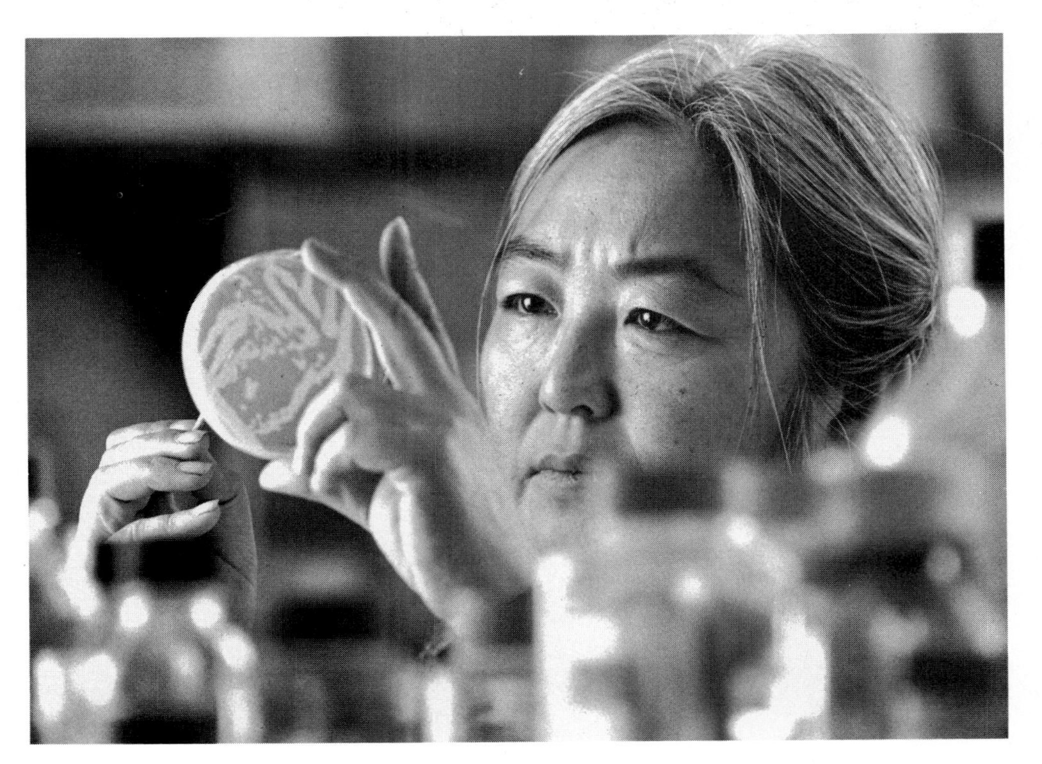

Dr. David Baltimore receiving the Nobel Prize, Stockholm, 1975 (AP/Wide World)

■

Baltimore at Rockefeller University (Marilyn K. Yee/NYT Pictures)

Walter Stewart (*left*) and Dr. Ned Feder (*right*)
(© 1993 Annie Adjchavanich)

Margot O'Toole testifying before the Oversight and Investigation Subcommittee. Walter Stewart (*middle*) and Dr. Ned Feder (*right*). (AP/Wide World)

■

Congressman John Dingell, Chairman of the House Energy & Commerce Committee (Philippe Jenney/*Legal Times*)

Peter Stockton
(© 1993 Annie
Adjchavanich)

Dr. David Weaver
(Ken Heinen)

Dr. Hugh McDevitt
(Ken Heinen)

Dr. Joseph Davie
(Ken Heinen)

Dr. James Wyngaarden, former Director of the NIH (Ken Heinen)

Dr. Ursula Storb (Ken Heinen)

Suzanne Hadley
(© 1993 Annie
Adjchavanich)

■

Bernadine Healy

tion . . . she had obtained indirectly and secretly from the laboratory notebook of another postdoctoral fellow."

Testifying along the same lines as Flax, Huber said that she and Henry Wortis had no conflict of interest in reviewing O'Toole's concerns. The Tufts scientists, more than anyone, would want to get at the truth, said Huber: "Why should we have wanted to recruit as a colleague someone who commits fraud and whose science is not trustworthy?"

Wortis, who had been O'Toole's thesis adviser, said that the ad hoc committee asked for and saw the data on Table 2 clones. (He was apparently referring to the June subcloning data that had changed the minds of the scientists on the NIH review committee.)

"It is important to realize that we did not consider ourselves a formal review committee conducting an official review," Wortis said. "We were acting as scientists to review a dispute among scientists as to the conclusions of a paper. There were at that time no allegations of fraud or misconduct."

The possibility of fraud, however, had been on their minds even if O'Toole would not charge it. "We specifically discussed that with her on more than one occasion, and she assured us that she was not accusing Dr. Imanishi-Kari of fraud but rather was questioning certain of the conclusions of the *Cell* paper," Wortis testified.

Wortis noted that there had now been three reviews of the paper and not one had found fraud or misconduct. Indeed, he said, the reviews stated that the "original conclusions of the paper are legitimate." And Wortis said he was unaware of any published experiments contradicting the central conclusion.

"Today we are engaged in yet another inquiry, this time by nonscientists, into the same matter," he lectured the lawmakers. "One can only ask, would it not have been cheaper and more compelling to hire a team of independent scientists to duplicate the study?

"Instead we have members of Congress of the United States apparently being asked to take sides in a scientific dispute. I am saddened and troubled by this process as it seems it is not truly directed toward the resolution of any scientific dispute."

Representative Lent probed the extent to which Wortis thought the dispute over the paper was the result of a personality conflict between Imanishi-Kari and O'Toole. Wortis said there was initially "a genuine scientific dispute," but "the fact that it was brought immediately outside of the laboratory may have been the result of the personalities involved."

Lent also fed Wortis a question about Tufts's interest in fully investigating the charges. Even though Tufts planned to hire Imanishi-Kari, it had not done so at that point, so did not Wortis have a motive to try to determine the truth of the allegations?

"That is true," Wortis said. "If we found that by omission, by sloppy science, without regard to intent, that she had produced results that were not supported, that . . . did not support the conclusions of the paper, we would have had doubts about our wisdom in hiring her."

Democrat Walgren was tougher. He wondered why no one had tried to confirm the experiments that O'Toole had challenged. "Do we not want to challenge a Nobel scientist in some way?"

Even though Wortis had earlier conceded to Dingell that the central conclusion, dealing with the mouse's own genes making the transgene idiotype, had not been confirmed in published experiments, he answered Walgren differently.

"In fact, the central conclusions have been confirmed. Those have not yet been submitted for publication, however," Wortis said to Walgren. He never gave details identifying this work.

Continuing to respond to Walgren's question, Wortis said that basically the *Cell* paper had been so convincing that other immunologists did not see any need to personally replicate the findings.

The subcommittee, Walter Stewart, and Ned Feder also had to consider what impact their efforts in making this case public had on why scientists wanted little to do with the experiments discussed in the *Cell* paper, Wortis said. "Has that led scientists to decide to stay away from this issue for fear that they might wind up spending three years enmeshed in trying to defend themselves when in fact they might not have anything that needed defending other than the science itself? That is a question that you can better answer than I," Wortis said.

Wortis fenced with Dingell and Walgren over the paper's incorrect statement that isotyping had been done on the Table 2 hybridomas and over the question of why his group had not discovered any serious errors while the NIH panel had. Wortis impatiently said that the isotyping had been done for Table 3, so the Table 2 claim was unimportant, and he pointed out that NIH had not found the conclusions of the paper to be invalid. He also noted that O'Toole had not initially raised the issue of missing data for Figure 1—she did not know of the missing data at the time.

But what about the corrections that the co-authors had to submit? Dingell asked.

"You've now talked about error, which is different from challenging the conclusions," Wortis said. "Well, I think we're going to go round and round on this because if there is a belief here that science in order to be correct has to be absolutely free of error, then we're stuck."

Dingell drew himself up. "I'm not so foolish as to say that it has to be absolutely correct. I'm saying that it has to be procedurally correct. It has to be factual. After all, I keep hearing—and maybe I'm in error or maybe the people who would say this are in error—that science is essentially a search for the truth. Is that a fair description of what science is?"

"Yes," Wortis said. "And I think the conclusions of this paper have been so far found to be true. So—"

But Dingell brought the discussion back to his question: "Well, why was it that the NIH panel suggested corrections and your panel did not?"

Wortis didn't give too much ground. "I think that what we're dealing with is statements of what I would call clarifications, and I don't—I don't disagree with the clarifications that have been made."

What about the suspicious dating of the subcloning data records? The Secret Service had reported that the pages were done after O'Toole challenged the paper and just before they were shown to Wortis's group. Was Wortis told when the data had been generated?

"No, at this time we were interested in the data," Wortis said. "I don't remember what the dates were on those pages, not at all. . . . The question is when were the experiments done, and I don't know if anyone has testified as to when those experiments were actually done. I don't think the Secret Service did that."

Dingell agreed that no one knew when the experiments were done. "We do have an idea what the dates were on the papers on which they were purported to have been recorded. We have sworn testimony from the Secret Service that the dates on the papers were changed."

Wortis's colleague Huber jumped in to say that it would have been impossible for Imanishi-Kari to have generated *in one week* the subcloning data to respond to O'Toole's challenge. "It's technically not possible." But she did say that, while the ad hoc panel had examined the subcloning data, they had not noticed the dates.

Dingell zeroed in: Did Huber not tell the subcommittee staff that at the first meeting with Imanishi-Kari, she had reviewed only the data associated with the mouse that had been mistyped as a control? "Is that true or false?"

"That is false," Huber said. "That was our main concern—was certainly my main concern."

"So, you're telling me that you did not tell the subcommittee staff that?"

Huber: "I'm not sure what I said in that meeting. . . . We did look at subclone data in our first meeting with Dr. Woodland—Dr. Wortis and myself were present."

Wortis joined in "because I can remember this very specifically." He recalled how O'Toole had questioned the presence of some transgene product in the Table 2 wells, and Imanishi-Kari said that the wells had not yet been cloned.

Dingell: "Does that mean that the data had been generated at the time?"

Wortis: "No, you can't do it then. Let me finish. So, we said yes, but what's important is whether those particular wells have been cloned." Imanishi-Kari said she had done so, and when they asked to see the data, she cried, Wortis testified. But she got the data out for their examination.

"Well, maybe you can tell me when you became first aware of the existence of the June subcloning data?" Dingell asked. "Did that occur at the meeting referred to?"

Inexplicably, Wortis responded, "I don't know which June subcloning data you're talking about."

■

Another House floor vote was called then, and when the panel reconvened Dingell did not return to the subject of the subcloning data. Instead, the panel turned to John Deutch, provost and chief academic officer of MIT, and Herman Eisen.

Deutch, a former under secretary of the U.S. Department of Energy who had both jousted and cooperated with Dingell before, was conciliatory in his opening statement.

"We recognize the importance of being and being perceived to be responsible custodians of public funds," Deutch said. "It would be a tremendous loss for this nation if misunderstandings about how science functions would lead to a lessening of congressional confidence or support for new regulations which might impair the scientific enterprise."

Representative Lent first asked Eisen if an analysis of the accuracy of the *Cell* paper could be done using seventeen pages from one lab notebook?

Eisen's answer was an emphatic "Most certainly not." Any selection "could have all kinds of misleading material in it."

While he did not look at the underlying data, Eisen said his review was not cursory. "I think I'd be in a pretty good position to have suspicions raised about fraud or misrepresentation of data, by carrying out the kind of inquiry I did carry out."

Eisen picked up the refrain of the supporters of Imanishi-Kari and Baltimore: O'Toole's charges, he said, kept changing. "What seems to be happening to me is that there is a changing nature of the charges and complaints being raised. That makes it sort of difficult to know what a termination or end point is. I think if the original issue was a paper that appeared in *Cell* . . . , I think the inquiries have done the job. . . . The paper itself is in the public domain. It is being evaluated in a way that is traditional and very effective scientifically and that is it."

In response to Dingell, Eisen did concede that the possibility of fraud was on his mind when he heard out O'Toole. But he noted that O'Toole said she was only concerned with error.

If Eisen saw the possibility of fraud, Dingell asked, why was MIT's policy on investigating allegations of fraud not triggered even though O'Toole did not specifically allege it? Why wasn't NIH informed of her allegations then?

Eisen blamed the lack of a fraud investigation on O'Toole, for firmly stating that she saw scientific error rather than fraud. Besides, he said, he had carried out the first phase of MIT's misconduct procedures by performing a preliminary investigation, which had not found sufficient grounds to go further.

"Is it fair to expect a postdoctoral student, a junior scientist, or even a senior scientist to charge fraud against a colleague before formal investigations are initiated at the school?" Dingell persisted.

Eisen: "Well, I think that there is a real point to what you're saying. That is to say, I could imagine real circumstances where a person who's bringing complaints has complaints of such a nature that without ever using the word 'fraud' there is real reason to perceive it as a charge of fraud. . . .

"I was not unaware of the possibility that she had in mind fraud and was unwilling to say so, and in carrying out my evaluation, this concerned me," Eisen said. That was why, Eisen said, he took so much time talking to people and thinking about the issues.

Dingell still wondered whether the "threshold someone must leap before they get a full and formal inquiry" might be too high. "It looks less like the doorsill in this committee room than it looks like the Great Wall of China."

When Deutch, the MIT provost, said that a preliminary inquiry was conducted, as called for by the university's rules, Dingell asked Eisen if he had looked at the laboratory notebooks in conducting his inquiry.

"I did not look at notebooks," Eisen said.

Did he know that there were experiments that were not done?

"No, I did not know that and I couldn't have known that and in fact, the question was never raised. That whole question of experiments not having been done was not raised by Dr. O'Toole then or until very recently so far as I know," Eisen said. (In fact, O'Toole's memo to Eisen had noted that some experiments had not been carried out.)

In any event, Eisen said, his was only a preliminary inquiry, which he did not think required an examination of all of the notebooks.

"But," Dingell said, "a formal inquiry would have gotten you through the notebooks."

Eisen wasn't going to get away without answering questions about Baltimore's September 1986 letter to him about Imanishi-Kari and the Bet-1 reagent. However, he had little more to say on the subject than Baltimore had had. He described his miscommunication with Imanishi-Kari as "one of those corridor conversations where people say things in passing. . . . When I got back to Dr. Imanishi-Kari, she explained much more about it, and I think what emerged from that explanation was the picture I still have to this day and we all have now." That picture, he said, was of a reagent that distinguished one gene better than another and that when made radioactive for use in assays would sometimes be damaged and become useless.

Dingell asked Eisen about his August 4, 1986, notes of a conversation with David Weaver in which both men agreed that a correction noting that Bet-1 was not absolutely specific as stated in the paper should be submitted to *Cell*. Why wasn't it submitted? Who opposed publication? If both Eisen and Weaver agreed to send the correction, only Baltimore or Imanishi-Kari could have raised an objection, and it would have been difficult for Imanishi-Kari to oppose Baltimore if he supported such a correction.

Eisen danced around his answer. He claimed that it was a minor point, and that anyway most scientists already knew or would soon know about the problems with the reagent. "But I didn't push it. I think it would have been maybe wiser to have pushed it in a political sense, but I didn't push it because it's a very minor point," Eisen said.

"Who was it that didn't agree?" asked Dingell again. "You said you didn't push it. I'm trying to understand with whom did you have to push it. . . . Dr. Weaver agreed, you agreed, somebody didn't agree, so it wasn't done. Now, who else was it that disagreed?"

"Nobody said, 'No, you can't do it,' " Eisen said. "They just were obviously clearly unenthusiastic and didn't want to undertake it, and I didn't push it."

In response to a question from Dingell, Eisen said there was some sentiment expressed that a retraction of the paper or "an overall concession that the authors accepted the validity of [O'Toole's] complaints" could hurt Weaver's fledgling career even though he was not responsible for the questioned work. Despite this admission, Eisen still maintained that the reason the paper was not retracted was that the authors did not believe O'Toole's complaints were valid, except for what they considered the minor problem of Bet-1.

What about O'Toole's career? Dingell again asked. Eisen conceded that her career was adversely affected when her doubts were not confirmed; he wondered whether the situation might have been different had she been able to publish her memo to him in some form.

"The visible result of all this," Dingell said, "is that Dr. O'Toole is sitting in this room with her career in shambles and a number of other people are prospering mightily. As one who is concerned with basic and fundamental justice, I'm curious, where is the justice in that situation?"

Both Eisen and Deutch said that O'Toole was not barred from a career in science. Deutch said that if an MIT lab found that her qualifications met its needs, she could be hired at MIT.

Dingell wasn't buying. To Deutch, he said, "You and I are old friends and we've been in this room on the same side and different sides through a lot of fights. Just speaking now as my old friend, would you want to tell me that you'd want to change places with Dr. O'Toole?"

Deutch's answer was clear: "No, sir."

And Deutch conceded that blowing the whistle, raising questions about possible fraud, is "a very difficult and dangerous thing to do. And I've given it some thought, Mr. Chairman. I really don't think it is easily resolved."

■

The policies and procedures governing academic fraud at MIT and Tufts were similar. Both universities generally defined fraud as fabrication of data, theft of ideas, and plagiarism. MIT included deliberate interference with the integrity of the work of others, and Tufts the violation of regulations governing research.

Under both sets of rules, anyone with a good-faith belief that fraud had occurred was to report to the accused's supervisor. MIT added that the accuser was free to seek advice from another responsible person, such as a dean or one of the special assistants to the president of MIT.

The accused's supervisor would conduct a preliminary investigation to determine whether there was reason to suspect fraud.

Under the MIT rules, the supervisor would make a written request to the accused for a meeting to discuss the allegations; the accused would have the right to be accompanied by another member of the MIT community. If the supervisor still believed there might be fraud, the issue would be brought to the provost, who, after consulting with faculty officers, would appoint a fact-finder or a committee from outside the affected department or lab. A full hearing would be held. At the same time, the provost would inform the sponsor of the research—for example, NIH—of the situation. After the hearing, the provost would decide whether to take disciplinary action.

The Tufts rules called for the accused's supervisor to report to the appropriate dean, who would determine whether there was probable evidence of misconduct, and if so, would refer the matter to the provost's Standing University Committee. If the committee concurred, it advised the provost, and disciplinary proceedings were started.

MIT also required "reasonable steps to protect those who have honorably raised concerns about fraud," while the Tufts rules specifically prohibited discrimination or retaliation against a person reporting misconduct or against any witnesses.

MIT's rules were in place at the time of O'Toole's complaint; the Tufts policy was not implemented until October 1988. Louis Lasagna, dean of Tufts's Sackler School of Graduate Biomedical Sciences, chaired a committee in July 1986 to draft a university policy condemning research fraud and to develop procedures for reporting and investigating allegations.

Lasagna testified at the House subcommittee hearing that in his five years as dean, no case other than O'Toole's had been brought to his attention. And the procedures, in effect only since October 1988, had not been tested by a specific case.

He noted that the procedures did not provide for a situation like O'Toole's challenge, because Imanishi-Kari was not then a member of the Tufts faculty. "We would probably have to have slightly different procedures for dealing with them, but I suspect we do need those," Lasagna said.

But Deutch insisted that MIT's misconduct policy had been applied to O'Toole's complaint—Eisen was asked to conduct a preliminary investigation—and that her concerns were treated seriously even though she did not charge fraud. "We do not require anybody suspecting this conduct to assume the burden of formally alleging misconduct before proceeding," Deutch said.

Deutch did note, however, that "on the basis of information currently available to us" from the Eisen and NIH reviews, neither [MIT president Paul Gray] nor I see any reason to believe that academic fraud was committed in this case."

The nearly six-hour hearing, which began with Dingell's praise for Margot O'Toole, ended with a harsh assessment by Norman Lent, who said that O'Toole had received the attention of "almost a score of eminent scientists" who "reviewed her complaints and found them wanting." He was impressed with the "thoroughness" of the inquiries.

Lent also noted that it was Henry Wortis who had gotten O'Toole the job in Imanishi-Kari's lab. "And then for Dr. O'Toole, being a student, to go through her [Imanishi-Kari's] notes and copy them and then turn them over to a man who was not qualified, who was not an immunologist, in the hopes that he would somehow bring this whole process into review, frankly, I'm a little disturbed by that kind of conduct," Lent said. "I know if I had someone in my office who did that sort of thing with my notes, they'd be out of there in a flash and they wouldn't be rehired by anybody that I could call up."

Dingell, who already had been swamped by letters from outraged scientists stirred up by letters from Baltimore and Phillip Sharp, and hit by criticism in the press since the 1988 hearing, would soon get new media reviews. In general, they would be as harsh as Lent's comments about O'Toole.

THIRTEEN

The War in the Press

■

David Baltimore's confrontation with John Dingell and his national campaign to whip up scientists against the oversight subcommittee resulted in a tremendous public-relations coup following his testimony at the May 4 congressional hearing. The editorial accounts of the panel's hearing overwhelmingly sided with the Nobel laureate.

Most of the press held its editorial fire until after the May 9 hearing. Margot O'Toole testified again; also included were the scientists from MIT and Tufts who had originally reviewed her complaints and decided there were no substantive problems with the *Cell* paper. *The Wall Street Journal*, which consistently defended Baltimore and attacked Dingell throughout the controversy, on May 5 ran an installment of Paul A. Gigot's "Potomac Watch" column titled "Latest Chapter in the Fine Science of the Smear."

The column discussed the nightmare lived by a scientist who, after leading an exemplary and accomplished life, wakes up and reads in his local paper that he's been accused of fraud. "His experience since has taken him into the Kafka maw of 'congressional oversight,' Dingell-Style," Gigot wrote. He referred to Dingell as "Congress's grand inquisitor" and to O'Toole as "a research associate (a 'whistle blower,' in Dingell parlance)."

Gigot seemed to go further than even Baltimore would, at least publicly: Congress had no right whatsoever to get involved. "Even if some 'misconduct' were proved, the salient question would remain why the U.S. Congress should care. Good science requires freedom, especially the freedom to blunder. . . ." Gigot wrote. "Mr. Dingell says he'd never, ever seek to 'police science.' But then why did he call in the Secret Service (a.k.a., the Treasury police)?" He concluded, of Dingell and his investigators, "The arrogance bred of unchallenged power has stripped them of self-restraint and distorted their understanding of the public good."

On May 15, *The Wall Street Journal* struck again with an unsigned editorial: "The Science Police." The editorial contended that Dingell was harassing scientists and "taking steps that would police science and eventually cripple it. He'll succeed unless science fights back." The editorialist was disdainful of Stewart and Feder. "While the Dingell committee has championed and publicized their efforts, others at NIH privately call what is going on the 'lionization of the turkeys.' . . .

"Important science is self-correcting. Plagiarists and con men eventually are discovered, which is more than one can say about Congress," the editorialist wrote, concluding: "David Baltimore's travail is only the beginning if scientists remain silent and let John Dingell become the Auditor General of American science."

An Associated Press article sent out on the wire the day after Baltimore's testimony was not an editorial, but was framed in such a way that it clearly was a plus for the scientist's campaign. The story led with a letter, distributed at the hearing, from a prominent pediatrician critical of Dingell.

"A noted pediatrician says his patients will suffer and die because of attacks by a powerful congressman on the work of a Nobel Prize–winning researcher," the AP article began. The piece quoted an April 26 letter from Dr. David Nathan, physician-in-chief at Children's Hospital in Boston and a pediatrics professor at Harvard Medical School: "An attack on him is an attack on our patients. . . . The David Baltimores of this world have made it possible for them to live. It is as simple as that."

Adding to the barrage was a piece in the May 14 *New York Times* about scientists' fears of congressional intrusion into science. The column, by reporter Warren E. Leary, said: "Visions of 'science police' making unannounced laboratory visits, pawing through notebooks, auditing data, and demanding explanations for every error are sending chills through the corridors of universities and research institutions nationwide." Baltimore's friend Phillip Sharp was quoted as saying that the attempt to associate

Baltimore with scientific misconduct was not "a search for truth, it's a witch hunt."

Science, the magazine of the American Association for the Advancement of Science, said in its May 12 issue that Dingell had done "his level best to pillory Nobel laureate David Baltimore [the previous] week. His principal stratagem: to catch Thereza Imanishi-Kari at fraud and watch her drag Baltimore down with her. He succeeded in neither count." *Science* reporter Barbara J. Culliton wrote that the expectation that the Secret Service had found "a smoking gun" had come to naught, although there was evidence that the preparation of the paper was "sloppy and even flawed."

Even Dingell's hometown newspaper, *The Detroit News*, attacked the lawmaker for going on an "apparent wild goose chase looking for 'scientific fraud,' " when he could have been doing something meaningful for his constituents, such as opposing new fuel-economy rules. In a May 17, 1989, editorial titled "Dingell's New Galileo Trial," the *News* presented Baltimore as a beleaguered Nobel Prize–winner with a "glittering resume" who was unfairly suffering from the "truly frightening power that John Dingell possesses. . . . Chances are, [Baltimore's] Nobel Prize will likely shield him from ruin, but one can only shudder at the consequences for those not so fortunate." The newspaper also blasted Peter Stockton, Dingell's aide, for reportedly accusing Baltimore of fraud and misrepresentation. The *News* concluded that the lawmaker could "do the American auto industry and his constituents a world of good if he would use his power to limit destructive [fuel-economy] regulations and leave the molecular biology to people who know and care about it."

Dingell personally responded to the editorial and refused to back off the case. He defended his subcommittee inquiry by again insisting that its purpose was to examine how research institutions "handle allegations of error, misconduct, and fraud in federally funded research."

The *News* editorial, he said, "simply parrots the line of the 'outraged' scientific community, which somehow feels threatened when its actions are examined. I note that *The Detroit News* did not send a reporter to the hearing and did not run news stories on the facts in this case. Perhaps that is why the editorial writers do not understand the issues."

Upon hearing the Secret Service evidence, the NIH had reopened its investigation, Dingell noted. The agency's response, he said, "has been measured and reasonable. The scientific community, on the other hand, had declared itself outraged, but has not bothered to inform itself of the facts. This is precisely the opposite of the scientific approach: to develop evidence adequate to resolve the contested issues as matters of fact."

Dingell also defended Stockton, saying that the text of the controversial *Boston Globe* article made it clear that his aide had not accused Baltimore of fraud. "There is an old saying in the law: 'When the facts and the law are against you, attack the integrity of the other side,' " Dingell wrote back. "The co-authors are trying to divert attention away from the facts and on to the conduct of my staff. This is the oldest trick in the book. . . . This supposed 'accusation of fraud' is just another strawman erected to divert attention from the focus of the hearing. The focus should remain on the integrity of the research project."

Although Dingell got quite a drubbing in the press, not all of the news media simply blasted the lawmaker and cast an uncritical eye on the scientific community. Daniel S. Greenberg, the publisher of the independent newsletter *Science & Government Report* and an early skeptic concerning Baltimore's side of the case, was more pointedly and harshly critical of the scientist and his supporters than most. He argued that despite winning the public-relations war, Baltimore and his supporters actually came off quite badly during the hearings. In the May 15 issue, Greenberg recounted the story of O'Toole's questions and the various inquiries and said that "despite the personality factors and the technical complexities of the case, the basic issue remains unchanged: How did the powers of science respond to the allegations of error? The answer, in one word: disgracefully." Greenberg's June 1 issue, which included embarrassing excerpts from the hearings, was headlined "Baltimore Hearing (Part II): A Torrent of Thin Alibis."

In a May 22 column printed in the *Baltimore Sun*, Greenberg noted that as far as press clippings were concerned, the Nobel laureate was "a runaway winner in his tortuous dispute with the powerful and irascible" John Dingell. "But the victory came from tactics that are no credit to science, and the basic issue remains unchanged in this confrontation at the two-cultures gap: Can science be trusted to police itself?"

Although O'Toole's challenge was justified, Greenberg wrote, "the science establishment responded with retribution, whitewash, and earnest assurances of upright self-governance in behalf of the public well-being." After ticking off the anti-Dingell headlines, Greenberg wrote, "Score a knockout for Dr. Baltimore. But the victory was produced by bare-knuckle, often disingenuous public-relations tactics remote from science's traditional ways of communicating with politics and the public."

A rare sympathetic treatment of Margot O'Toole's story that went beyond a simple confrontation of personalities ran the following month in *The Washington Times*. Written by Diana West, "Anatomy of a Scientific Scandal: O'Toole's Whistle" pointed out that the Secret Service findings supported O'Toole's contentions by showing that data purportedly prepared in 1984 and 1985 actually had been produced in 1986, after publication of the *Cell* paper.

■

While some observers thought Dingell was shaken by the confrontation with David Baltimore at the hearing, and while he had clearly lost the early public-relations contest, the lawmaker was philosophical about his setback, as a person who is truly self-confident can be. Describing himself as "an old infantryman," Dingell explained his tactics in military terms: He simply scouted the enemy, made a plan, carried it out boldly and furiously, and then consolidated. "You're going to suffer a number of losses," Dingell went on. "I've been here a long time in this trade. I'm not concerned with what happens that day, that year, that session, that decade." If the issue was important, Dingell would outlast his adversaries.

Baltimore might have won the battle, but unbeknownst to his jubilant supporters at the congressional hearing and to casual readers of the newspaper headlines, he was already starting to lose the war.

Some influential friends of Baltimore's at Harvard were becoming dismayed with his defense of the *Cell* paper, which they increasingly believed could not be defended. One of those friends, who later became considered a member of what Baltimore and Imanishi-Kari supporters would call the Harvard Mafia, was Mark Ptashne.

Ptashne had gone to Capitol Hill for the May 4 subcommittee meeting to hear the testimony for himself. Although his impression of Walter Stewart from Cold Spring Harbor was—to put it mildly—poor, Ptashne was not satisfied with the answers that he was getting about the *Cell* paper. He got hold of the five papers that Baltimore implied confirmed his own and read them over the weekend.

During a break in the hearing, he introduced himself to O'Toole and told her, "You're right; they didn't repeat it." She was stunned: A senior scientist was supporting her. Ptashne, who once worked in the Harvard lab of Nobel laureate Walter Gilbert, was a noted scientist in his own right. He also, like Baltimore, had a reputation for arrogance. As O'Toole stared dumbly at him, Ptashne wondered aloud about what could be wrong with the *Cell* paper.

Ptashne was still trying to figure out the *Cell* paper and reconcile his questions about what he thought was a sloppy paper. "I don't know if everything Margot O'Toole says is true. I don't know if she's a great scientist. But what she said made sense," he recalled. "She sure as hell knows what's going on."

After the hearing, Ptashne would call O'Toole for ten-minute snatches of asking questions about the *Cell* experiments. During one call, he told her: "I've talked to everybody. You're the only one who's acted like a scientist."

Several weeks later, trying to keep the shakiness out of her voice, O'Toole called Ptashne to ask for help in getting a job. He didn't seem to understand the problem.

"It's been three years and I have no job," she explained.

"I'm sure it'll all come out that you're right," he told her.

"But I need a job now," O'Toole persisted, in spite of her nervousness.

Ptashne did not make any promises before hanging up. However, he recommended O'Toole to the Genetics Institute, a biopharmaceutical firm that he cofounded but was only an adviser at that time. The Cambridge company, which is in the business of discovering and developing new pharmaceuticals through recombinant DNA, called her shortly afterward, although she was not actually hired until the following April, as a temporary technician. (The ten-week job satisfied her because O'Toole was seven and a half months pregnant then with her third child. Genetics Institute later hired O'Toole, the former telephone receptionist for her brother's Gentle Giant Moving Company, as a full-time staff scientist.)

What Ptashne and O'Toole did not know immediately after the 1989 hearing was that there were other prominent Harvard scientists whose doubts about the *Cell* paper were growing. One of those skeptics was Wally Gilbert, a molecular biologist who shared the 1980 Nobel Prize in chemistry with Paul Berg and Frederick Sanger for his work in developing a technique for mapping the DNA code, the sequence of chemicals that make up the DNA strand. Gilbert also discovered how to induce bacteria to produce insulin, which can then be harvested for use by diabetics.

Gilbert was a longtime friend of Baltimore's and Phillip Sharp's. He had known Baltimore since about 1968; their families sailed together in the South Pacific. Gilbert and Sharp had worked together to form the biotech company Biogen, Inc., in 1978.

When the *Cell* paper came out in 1986, Gilbert thought the idiotypic mimicry conclusion was exciting. "I accepted the arguments," he recalled. He also started hearing through the grapevine that complaints about the paper were being made by a disaffected postdoc. When Gilbert received a copy of the Stewart-Feder paper, he did not think much of it; he knew about them through their work on the Darsee case. While Gilbert thought their Darsee paper was exaggerated, he still felt that Stewart and Feder were "on the right side" of the scientific-integrity issue.

But as Gilbert started learning more about the *Cell* paper, he came to believe that its conclusions were incorrect; the Bet-1 reagent did not work properly. Gilbert read O'Toole's testimony from the first 1988 congressional hearing and was impressed. The scientists who were saying the argument was over interpretations were wrong, Gilbert decided. He understood and believed O'Toole when she said the issue was one of fact: "Does the paper reflect what's in the notebook?" It did not seem to Gilbert that anyone besides O'Toole was really looking at the underlying data.

The pivotal point to Gilbert, however, and apparently to a number of other scientists as well, was Imanishi-Kari's testimony at the 1989 hearing, when she argued that she had no motivation to fake her work because of its potential for leading to a cure for lupus, the disease that struck her and killed her sister. What had played sympathetically in the news media was a warning to Gilbert that something was seriously wrong with the *Cell* paper. "That's a bad motivation for science," Gilbert said. "It means you are more likely to fake [the results]."

Gilbert was also distressed by Baltimore's campaign against Dingell. "I felt he was wildly wrong in doing that."

Ptashne and Gilbert had not been particularly close for some time, so there had been no reason to discuss the case. But one day in passing, Ptashne threw out the question: What did Gilbert think about the *Cell* paper? Fraud, Gilbert reportedly replied, and moved on.

The outcry over Dingell's subcommittee investigation continued through the summer. In a July 30 op-ed piece in *The New York Times*, Harvard geology professor Stephen Jay Gould compared the controversy to Galileo's conviction for heresy for teaching the Copernican system. Under the headline "The Perils of Official Hostility to Scientific Error," Gould wrote that Galileo has been considered one of the great scientific martyrs but "in light of recent developments in Washington, I'm not so sure that Galileo might not be in more trouble today."

Gould argued that the *Cell* paper contained errors, not fraud. While scientists must accept some scrutiny from the people who fund their work, "what could possibly be more chilling to creativity than an office of censorship?"

Baltimore himself continued to speak out and write about the case. In the August/September issue of *Technology Issues Review*, Baltimore maintained that O'Toole had simply disagreed with the interpretation of the *Cell* paper's co-authors. As for the Secret Service, he said the agents "could analyze only paper and ink, not data and its interpretation." He contended that in his confrontation with Dingell he had fought not only for himself, but for all scientists.

"Such government intervention in the scientific process is potentially destructive. If this type of investigation had not been countered strongly, its threat would hang over all researchers causing fear and regimentation and ultimately, stifling creativity," Baltimore wrote. "If each piece of science has to be so cut and dried that a congressional investigator can audit it, then research will become a defensive process. Too much energy will go into preparing for the audit rather than in grappling with the unknown. . . . The battle I have been fighting is for my reputation and that of my colleagues."

Despite the generally favorable notices that the Baltimore campaign received shortly after the congressional hearings, and the scant public support for Margot O'Toole, there was a growing uneasiness within the scientific community over the Nobel laureate's actions—and not just at Harvard.

■

Baltimore's name surfaced in September on the list of leading candidates to replace retiring Rockefeller University president Joshua Lederberg. From the beginning, there was significant opposition to Baltimore because of the way he had handled the controversy; some faculty members thought it reflected badly on his ability to lead the proud university. Others thought that a continuing controversy about Baltimore, no matter that his own work had not been questioned, would nonetheless leave a cloud over Rockefeller. A straw poll of the forty-five senior faculty members indicated that only a third supported his candidacy. Some of the senior scientists who opposed his appointment had been his classmates in the 1960s at Rockefeller, where he had earned his doctorate.

The university, located on York Avenue at 68th Street in Manhattan, was founded in 1901 by John D. Rockefeller, Sr. Ranked among the top biomedical-research institutions in the world, Rockefeller University enrolls only graduate students, and they are far outnumbered by the faculty, postdoctoral students, and research associates. Baltimore was strongly supported for the presidency by David Rockefeller, the former chairman of the Chase Manhattan Bank, who was head of the trustees' executive committee, and by Richard Furlaud, president of the Bristol Myers Squibb Company, who was chairman of the board of trustees. Another of his strong supporters among the trustees was Stanford's Paul Berg, a Nobel laureate and Baltimore's comrade-in-arms on ensuring safe experimentation with genetic engineering. The trustees wanted Baltimore to lead the university through a series of reforms to bring in and encourage younger faculty members, help cut operating deficits, and generally give the research institute a new sense of mission. If Baltimore could build the Whitehead Institute from scratch into a preeminent biomedical-research institute, the trustees were sure he could certainly do wonders for Rockefeller.

In the face of the faculty opposition, however, Baltimore withdrew his name from consideration in late September. At that point, according to *Science*, David Rockefeller and Richard Furlaud flew to Cambridge to persuade him to change his mind and take the post. Furlaud quoted Rockefeller as telling Baltimore: "Look, we still think you're the right person to do the job." On September 29, faculty members found notices in their mailboxes that Baltimore had been offered the job and had agreed to consider it.

The trustees had looked into the controversy "in considerable depth," Rockefeller said, and found nothing "which could possibly reflect on the integrity or honesty or honorability of David Baltimore."

But the faculty members who had opposed Baltimore were not placated. Norton Zinder, a member of the faculty search committee advising the trustees and a supporter of Baltimore's at the Dingell hearing, said the trustees had not helped the situation by moving ahead on the appointment without the faculty's agreement. Baltimore wasn't even Zinder's leading candidate.

Anthony Cerami, a professor of medical biochemistry and dean of graduate and postgraduate studies, and Gerald Edelman, a Nobel Prize–winning molecular biologist, were among Baltimore's critics at Rockefeller. They thought his confrontational reaction to Dingell had not helped science, but instead had increased the likelihood of political interference.

Baltimore made things worse "for himself and science in general," Cerami said. "Unless someone comes in who can lead the faculty and gain its respect, it will be warfare for the next twenty years. It will be an unpleasant place to work in."

The trustees—and Baltimore himself—lobbied his critics on the Manhattan campus. While the public criticism had been embarrassing, Baltimore said he had cordial discussions with the scientists. "I very much feel bygones should be bygones. And I feel we should sit down and work for the future. I honestly believe their goals and my goals for the university are quite similar," he said in an interview with *The New York Times*.

In the midst of the dispute on the Rockefeller campus, *The New York Times* gently criticized Baltimore in an unsigned editorial titled "Ruckus at Rockefeller," saying that he had not pursued O'Toole's allegations "with appropriate assiduity." The *Times* said that Baltimore was "surely qualified" for the post, although it noted that his friends "have blown up and confused a rather specific issue by waving the flag of Galileo and portraying Mr. Dingell's investigation as a benighted attack on science. In fact,

Mr. Dingell is just trying to sort out a tangled case, and even the National Institutes of Health has recently reopened its inquiry." The *Times* concluded, "Universities are still trying to devise ways of dealing with disputed research and Dr. Baltimore did not commit the crime of the century in mishandling the inquiry into this particular case. . . . But the trustees took a risk in extending the invitation before the dust had settled. They should not be surprised that some faculty members wish to understand the case better."

When it became clear that the trustees were determined to make Baltimore president, Zinder and others urged their colleagues to rally around Baltimore in the interest of the university. "When the trustees had moved precipitately, we had no choice and we should close ranks behind David," Zinder said.

And on October 17, 1989, Rockefeller University announced that Baltimore would succeed Lederberg effective July 1. "For me, coming here is coming home," Baltimore told the Rockefeller community. He also defended his actions in dealing with Representative Dingell, contending that he had been left with little choice but to confront the lawmaker. "There are certainly events that in retrospect I might have dealt with differently. I believe that by confronting Chairman Dingell directly, I was acting in the interests of all scientists. Only time will tell."

The publicity over Baltimore's role in the controversy did not cease when his appointment was announced, however. An article in the October 29, 1989, *New York Times Magazine* about the case was headlined "Conduct Unbecoming?" The writer, Philip Weiss, said that at times the scientific community's handling of O'Toole's complaints "smacked of the old-boy network." Weiss also wrote that "the Rockefeller appointment revealed rifts within the scientific community, which had seemed to speak in one voice in support of Baltimore."

If some scientists were becoming increasingly critical of Baltimore's actions, his pending departure for Rockefeller also revealed the continuing bitterness of some scientists at MIT toward Dingell's inquiry. In a November faculty newsletter, MIT scientist Jonathan King said that the loss of Baltimore "thins the ranks of scientific leadership at MIT and deprives the institute of one of the most qualified candidates under consideration for MIT's presidency." The congressional hearings "probably were a factor in retarding Baltimore's candidacy here," said King, who charged that "Baltimore was made a scapegoat to an emerging effort to cut the growth of federal spending for biomedical research."

■

Dingell, however, did not let the scientists' charges of witch-hunts and McCarthyism deter his investigation. He—and the NIH—sent the Secret Service back to Cambridge to gather more information about the radiation-counter tapes.

FOURTEEN

Jousting

■

Following the May 4 and 9 Dingell hearings and the reopening of the NIH probe, Thereza Imanishi-Kari and her lawyer, Bruce Singal, carried on a lengthy debate with the investigators over whether and how she would cooperate with their inquiries. They jousted by mail, in telephone calls, at press conferences, and during a 1990 congressional hearing. Although anyone with a passing knowledge of the controversy knew that by that point Imanishi-Kari was being accused of making up experimental data, her lawyer wanted to know the formal allegations and review the documentation, including her original laboratory notebooks, before allowing his client to answer more questions.

On May 31, 1989, Brian W. Kimes, the acting director of the NIH's new Office of Scientific Integrity, wrote to Imanishi-Kari at Tufts, informing her that OSI was now responsible for the case and was focusing on an audit of the scientific data and on obtaining new information regarding the accuracy of the *Cell* paper. He asked particularly for any information about or analysis of the Secret Service findings.

More important, Kimes asked for Imanishi-Kari's help in developing an index of her lab notebooks, which would help explain what experiments were performed, and when. Without such an index, the investigators would be forced to make sense of the myriad counter tapes, handwritten notes, and dates on their own. He asked her to send the index and additional information by June 19.

Kimes received his reply on July 27, from Singal. Imanishi-Kari could not respond intelligently because she had not received any significant information relating to the Secret Service forensic studies. Without that information, Singal wrote, she could not "avoid unfair surprises of the type with which she was previously confronted when meeting with the subcommittee staff."

Singal said that his client must "be afforded the elemental right of receiving notice of the allegations" and copies of all documentation supporting them, because of the collaboration between NIH and the oversight subcommittee. "Because she has to date been deprived of these rights, regrettably I am unable at this time to agree to any NIH interview of Dr. Imanishi-Kari or to her submission of information to NIH," Singal wrote. "Dr. Imanishi-Kari is anxious to continue cooperating with NIH and looks forward to meeting with you once she has received and reviewed the necessary information."

Any contacts with the Tufts scientist, her lawyer added, should be routed through him so that she could concentrate on her science.

Kimes responded to Imanishi-Kari, however, on August 10, saying that OSI was "extremely disappointed" that she was not helping in the development of the index. He said that the OSI had not updated her earlier on the progress of the investigation because the office was first trying to formulate the allegations clearly and secure the subcommittee evidence. He also noted that she was not helping by refusing to develop the index.

Kimes tried to assure Imanishi-Kari that the OSI investigation was independent of the congressional probe, although it would take into account the subcommittee's findings. He added that she would "have every opportunity to see and comment on all information supplied to the NIH panel of experts." The panel of scientific experts also was being expanded; Kimes's letter did not identify the two new members, but they were Stewart Sell, of the University of Texas, and William McClure, of Carnegie-Mellon. Both were biologists.

Although the allegations were still being drawn up, Kimes said that the investigation would be considering questions about Figure 1, which was related to the Bet-1 reagent, and questions about new allegations concerning Table 2 data and the possible fabrication of subcloning data. Despite Singal's refusal to let Imanishi-Kari answer more questions, Kimes pressed on, asking for an account of how the *Cell* paper was drafted and inquiring what aspects had been confirmed by other scientists' published work.

"While we will move with as much dispatch as possible, our main priority is to reach a conclusion that will be final and not vulnerable to criticism," Kimes said.

Robert Lanman, the NIH legal adviser, also responded to Singal on August 14, saying that NIH did plan to make available all the allegations and the documentation behind them. However, he said, Singal's refusal to let Imanishi-Kari provide an index "has delayed implementation of these plans."

Furthermore, Lanman said, NIH would continue contacting Imanishi-Kari directly despite Singal's request to route inquiries through him. "It is entirely appropriate for NIH to communicate directly with individuals who are or have been supported with NIH funds," Lanman said, adding that he would send Singal a copy of each letter.

Singal complained to Kimes in an August 23 letter about the assertion that Imanishi-Kari and he were delaying the OSI investigation. "In fact, my client and I desire nothing more than that your investigation proceed promptly and that we assist and cooperate in it in every way possible," Singal wrote. "However, we are constrained from doing so by the apparent unwillingness of the congressional subcommittee to provide us with the fundamental information which is necessary for us to ensure that the two overlapping investigations are conducted fairly and in full observance of my client's rights." Singal asked for details about the new allegations.

Kimes told Singal in a September 14 letter that OSI was proceeding with its investigation without Imanishi-Kari's index, and that the office was in the process of drafting a written statement of allegations. He wrote Imanishi-Kari on October 5, asking that she confirm that the OSI now had a complete set of records for all the data relating directly to the *Cell* paper and those data that supported the results. The OSI was particularly concerned about "experiments conducted by you or Dr. Reis to isolate and characterize subclones of the Bet-1 hybridoma cell line and all experiments to examine the effects of iodination on the specificity of the Bet-1 reagent." He asked that Imanishi-Kari provide this confirmation by October 16.

The response came in a November 7 letter from Singal, who

was upset not only because Kimes's letter had been sent directly to Imanishi-Kari but also because he was not afforded the courtesy of a copy. Singal said that anyway his client could not respond to unfocused charges and the "unlimited scope" of Kimes's questions. "Your letter includes requests for information unrelated to the *Cell* paper. Please understand that Dr. Imanishi-Kari has worked with Bet-1 and transgenic mice for many years in numerous experiments unrelated to the *Cell* paper," Singal wrote. As for the complete lab records, Singal said those had been furnished long ago.

Additionally, Singal wrote, the OSI was basing its requests on "erroneous premises." Imanishi-Kari did not isolate or characterize the Bet-1 reagent, but instead relied on the published results of Dr. William Paul's lab at NIH. Imanishi-Kari had never conducted independent experiments on the effects of iodination on the reagent's specificity, Singal stated.

On January 10, 1990, Singal received an investigation update from Suzanne Hadley, now the acting director of the OSI, and Lanman. They said they were still working on clarifying the issues before the OSI investigation. However, they noted that the focus of the inquiry was changing from "questions about the accuracy of the reporting in the *Cell* paper to fundamental questions about the authenticity and integrity of the data that underlie and are related to the *Cell* paper." The investigators were also examining grant applications that appeared to be based on the same data as that used for the paper. The OSI again asked for a comprehensive index and other information. "Given the seriousness of these issues, cooperation by you and your client is all the more important. Indeed, we would think that your desire to be actively involved, to present your own perspective on the events and evidence related to the creation of the *Cell* paper, would be very strong," Hadley and Lanman wrote.

They apparently took into account Singal's criticism of the previous request and modified the new one to include data on "(a) specificity of Bet-1; (b) isolation and characterization of subclones reported in the *Cell* paper; and (c) effects of iodination on specificity of Bet-1."

Hadley and Lanman repeated the previous assurances that the OSI investigation was independent of Dingell's, and they made clear that refusal to cooperate would not stop the OSI investigation.

"You have suggested that your concerns about the manner in which Congressman Dingell's subcommittee has dealt with this matter are among the reasons you are not cooperating with the OSI investigation. But the OSI investigation is a separate undertaking, one conducted by scientists, and one that is devoted first and foremost to determining the truth. . . ," Hadley and Lanman wrote. "Without [Imanishi-Kari's] cooperation, we will be forced to rely on documentary and statistical evidence and testimony provided by others about the creation of the *Cell* paper. This may make the process a more difficult one for the OSI, but we certainly shall proceed with the investigation, with or without your client's cooperation."

Two days later, Hadley informed Imanishi-Kari that because she was a subject of a scientific-misconduct investigation, her name had been entered into the Public Health Service's alert system, which provides information about individuals charged with or found guilty of scientific misconduct to PHS officials responsible for making decisions on grants.

■

On February 1, Imanishi-Kari was informed by Hadley that she was formally the subject of the investigation, and that some of her co-authors might be added later. Hadley also described the focus of the investigation: "(1) the possibility that substantial portions of the claims related to the immunological aspects of the *Cell* paper were not supported by proper experiments and reliable data at the time of the paper's submission; (2) the possibility that after the problems with the paper were brought to light, there was systematic fabrication and falsification of data to support the paper; and (3) the possibility that falsified and fabricated data regarding immunological aspects of the paper were included in representations to the National Institutes of Health and in published letters of correction to the 1986 *Cell* paper."

A more detailed list of the issues was included, with the allegations linked to pages in the I-1 (Imanishi-Kari) and R-1 and R-2 (Moema Reis) notebooks. The OSI was clearly going after Tables 2 and 3, Figure 1, and the June subcloning data, all data that related directly to the central conclusion of the *Cell* paper or bolstered it after the fact.

Three weeks later, Lanman and Hadley sent Singal a copy of the charge given to the panel of scientists advising the OSI in the reopened investigation. The panel basically was directed to advise the OSI on whether scientific misconduct occurred in the "conduct, proposing, or reporting of the research," and, if so, who was responsible and how serious the misconduct was. Scientific misconduct, according to the Public Health Service, included "fabrication, falsification, plagiarism, or other practices that seriously deviate from those that are commonly accepted within the scientific community." Honest error or differences in interpretation are not considered misconduct.

Following a telephone conversation, Hadley again wrote to Singal on April 19, 1990, urging Imanishi-Kari's participation. "As I told you, the situation is one in which there is a mounting body of evidence that suggests there may be very serious problems with the authenticity of key sets of data associated with the Weaver et al. *Cell* (1986) paper, as well as the letters of correction and related letters and statements provided to the National Institutes of Health," she wrote. "Against this body of evidence, we have had no communication whatsoever from Dr. Imanishi-Kari."

Hadley warned that the investigators would soon begin to summarize their findings, but if Imanishi-Kari did not cooperate, her point of view would be absent from the report.

Singal objected in a May 1, 1990, letter to the implication in Hadley's letter that Imanishi-Kari had been shown a "mounting body of evidence" about which she refused to comment. Imanishi-Kari, he said, was still waiting for the charges and evidence against her before responding. A listing of the issues under investigation was not enough for Singal.

"We continue to wait patiently for you to present us with any such evidence which you believe incriminates Dr. Imanishi-Kari," Singal wrote. "Let me reiterate that if you do present us with a detailed recitation of the allegations against Dr. Imanishi-Kari, the body of evidence supporting these allegations, and the subject matters and issues to be covered at a meeting, Dr. Imanishi-Kari will be happy to meet with NIH-affiliated scientists to discuss her data. Until and unless you present her with any allegedly incriminating evidence, I would respectfully urge that you not create the false impression that you have already done so."

Hadley later noted in a May 10 letter that it was indeed true that Imanishi-Kari had not been presented with all of the evidence, so no inference should be drawn from the fact that she had not responded to it. However, she added: "At the same time, it is also true that Dr. Imanishi-Kari has failed to respond to straightforward requests for information, especially those posed and restated in our letters of May 31, 1989, August 10, 1989, and January 10, 1990." A detailed statement of the evidence would be provided before any interviews with the OSI staff, Hadley said.

What Singal referred to as allegations, Hadley apparently called "issues." She said in the letter that "a detailed statement of the issues that are the focus of the investigation was previously provided to you."

■

During this dance around the question of Imanishi-Kari's possible cooperation with the NIH investigation, Dingell was also trying to get the scientist to answer questions from his staffers and the Secret Service agents. Singal responded on July 27, 1989, to a request for an interview by complaining that Imanishi-Kari had been surprised at the last hearing by Secret Service information not provided at the earlier briefing and that it was unfair to question her without providing a written report of the forensic evidence. Before she would respond further, Singal said, she needed copies of all written forensic reports, exhibits, and other detailed information regarding the alleged alterations of her lab notebooks.

Singal also questioned the extensive cooperation between the subcommittee staff and the NIH. "My concern about the relationship is that it raises the potential of impugning the independence of the NIH review," Singal wrote. "We believe such independence to be critical to the ability of NIH to perform its duties in a fair and impartial fashion."

Dingell was not impressed. He said that the NIH had provided Imanishi-Kari with copies of some of the documents and exhibits from the hearing, but that it would not be appropriate to provide the forensic results of every test at that point in the investigation. He added that the Secret Service had testified fully about the agents' conclusions about their tests as of that point, and there was no written report yet. "The refusal of your client to cooperate with this investigation is wholly extraordinary . . ." Dingell said. "I find the unwillingness of an NIH grantee to cooperate fully with the NIH investigation and the subcommittee investigation deeply troubling."

The lawmaker wrote Singal again on January 26, 1990, in response to another request for a detailed statement of the nature of the investigation and the charges against Imanishi-Kari, as well as the questions the subcommittee wished to ask. Dingell sounded impatient: "The matter under investigation, as you and your client have repeatedly been informed, is the authenticity of the data submitted by Dr. Imanishi-Kari to the NIH panel of scientists and subsequently furnished to the subcommittee in response to its subpoena," Dingell said. "As the investigation relates to your client, the specific allegations being investigated are that she fabricated much of the notebook I-1 in an attempt to mislead the NIH and the subcommittee."

Dingell also was not accepting any excuses about Imanishi-Kari's language abilities: "Your statements about the inadequacy of Dr. Imanishi-Kari's English, and her inability to understand and answer questions, is not supported by any evidence available to the subcommittee. In fact, the evidence that is available contradicts your assertion, and shows that her ability to speak and to comprehend English is excellent."

Singal did not back down. In fact, he maintained that Dingell's suggestion that much of the I-1 notebook had been fabricated was "stunning." "Even the Secret Service testimony related only to a minuscule portion of the data contained in that notebook," Singal wrote back March 6. In the absence of more information from the Secret Service, Singal said, the agents' analysis "represents a sweeping vindication of the accuracy and integrity of the *Cell* paper. . . .

"Your framing of the allegations in a manner which employs draconian phraseology, smacking of suggestions of federal criminal violations, does a grave disservice to Dr. Imanishi-Kari and the other co-authors of the *Cell* article. . . . If such evidence truly exists—and I cannot believe that it does—then the time has come to share it with the accused," Singal said.

If Dingell replied specifically to this letter, he did not include his answer with the other letters and documents he had published in the subcommittee's hearing record. But on May 7, 1990, he asked Imanishi-Kari to testify at a subcommittee hearing scheduled seven days later. With his request he included copies of two Secret Service reports, one covering the results from the year before and the other based on new forensic studies.

Singal said in a May 10 letter that the reports were inadequate; he told Dingell that Imanishi-Kari declined the invitation to testify at the hearing. And she would not testify so until she received twenty-five documents or full explanations relating to the Secret Service findings, including the originals of the exhibits and of all notebooks from her laboratory.

Imanishi-Kari's lawyer also contended that the Secret Service agents had not questioned any pages that contained data published in the *Cell* paper, and that none of the cited pages included results material to the central findings.

While Imanishi-Kari would not answer questions or make a statement at the upcoming hearing, Singal asked Dingell to include his May 10 letter in the hearing record. But Singal and Imanishi-Kari were not going to rely on the inclusion of that letter as the only way of getting her views across to the public. Perhaps taking a lesson from David Baltimore's public-relations battle plan of the previous year, Imanishi-Kari also held a news conference in Washington on May 10 to discuss the Secret Service findings and attack the subcommittee. She, at least, would be able to present her perspective first, and the reporters certainly would be easier to talk to than John Dingell. However, after two years of stories and charges and countercharges, the reporters as a group seemed prepared to question Imanishi-Kari more directly than they had questioned Baltimore the previous year.

Singal started out the news conference calling the Secret Service reports a "sham," designed to create "an aura of fairness." He also cast Imanishi-Kari as a dedicated scientist, whose science was her life.

"I think I want to correct Bruce in one thing he said," Imanishi-Kari began. "He said that science is my life, my only thing in life. Actually science is part of my life. I have a kid that's also my life."

This year's Secret Service reports were puzzling to her: They did not seem to be saying anything new, and they did not relate to data published in *Cell,* she argued.

"As a scientist, my professional life is dedicated to reviewing and analyzing data. Yet the Dingell subcommittee do not give me any data to respond to," Imanishi-Kari said. "It seems that they are saying I made up this data. But then I would ask this question: Why would I fabricate data? . . . This data isn't in the paper, so why would I make it up?"

She commented further about the Secret Service work later in the news conference. "I believe Secret Service are scientists like me because what they do is science essentially. And I believe that as a scientist you have your results. You have your data. Like I have to give my books to them. . . . They should give me their data so that I am able to look [at what] they are talking about."

Peter Gosselin of *The Boston Globe* noted that the Secret Service report seemed to be saying that the radiation-counter tapes were not produced when the lab notebook claimed they were. Imanishi-Kari explained, as she had during her testimony in 1989, that the dates could mean different things.

"The dates on the pad can mean the date that the experiments were actually counted. Can mean the date experiment [began]. Can mean the date that the whole experiment was finished. So the date not always reflect the date of the actual printout coming out," she said.

But Gosselin pressed, asking about Singal's March 27, 1989, letter to Dingell certifying basically that "all of the data presented in the *Cell* paper were assembled contemporaneously with the scientific experiments, and placed in the notebooks in a timely fashion." Singal did not remember the letter, and another reporter quoted a portion of it.

"Right. Contemporaneous to the scientific experiment, which could mean around that time of when I started the cells, when I ended it up," Imanishi-Kari responded. That a page might have been written up after the experiment was unimportant, she explained. "You have to remember that pages is not experiment. . . . Now, a page doesn't say when a page was made, doesn't say when the experiment was done. Because the actual count is in the printout. The actual data is in the printout."

To another reporter, Imanishi-Kari disputed the contention that some of the questioned unpublished data had been used to defend the paper. "I didn't give anything to defend myself. I was requested to give all the books," she said. "They look at it because they wanted to look at it."

Gosselin also put a fine point on Imanishi-Kari's recording practices. "If you say that the date in your notebook cannot be interpreted to mean the date the experiment was done, do you have a record of when these experiments were done? . . . So you have no independent record of when you did the experiment?" he asked.

"No," Imanishi-Kari replied. She later said that "dates are irrelevant" in her notebooks.

A reporter from *Chemical and Engineering News* pointed out

that while Imanishi-Kari said that no problems had been found with the *Cell* paper itself, the NIH had required corrections, and some of the pages questioned by the Secret Service had been provided to bolster the paper.

Imanishi-Kari said that the reporter had misunderstood, because the corrections were minor.

The reporters also questioned why Imanishi-Kari was refusing to cooperate with the NIH investigation. Warren Leary of *The New York Times* asked: "So in other words you are not giving them certain information they have asked you for because you're waiting for them to give you something?"

Singal replied that the NIH had all the raw data, and the investigators "have expressed an interest in interviewing Dr. Imanishi-Kari."

"And you're refusing to permit that interview?" asked Gosselin of the *Globe*.

"No," Singal said. "We're agreeing to permit it, just as soon as they provide us with some basic information which we have a right to have."

This time, Greg Gordon of *The Detroit News* jumped in: "Well, isn't that a little unusual, if there's an investigation going on by a government agency, that you would demand information from them before giving the interview?"

Singal went through his explanation that his client needed documented allegations before she could respond in any meaningful way. Gosselin said more documentation probably wouldn't help resolve any questions because Imanishi-Kari's response would be that she didn't have the dates of the experiments.

Sheila Hershow of ABC asked the most direct question of Imanishi-Kari: "Did you fake the data?"

"No," Imanishi-Kari replied.

Also in response to reporters' questions, Imanishi-Kari said that she was penniless and that Singal was working pro bono.

In response to questions about whether the central conclusion of the *Cell* paper had been replicated, Imanishi-Kari seemed to back down on just what the central finding had been. "The central point of the paper is that in the transgenic mice, when you put this foreign gene, you alter the host gene product. . . . And that these host gene products look like the product of the gene you put in," she said. "So essentially is that something is changed in the mouse in terms of the host antibodies. This was replicated by many labs. . . . The second part of the issue is that the host antibody has the idiotype. . . . A similar idiotype as the gene produced has been replicated."

She said that she basically had replicated the idiotype finding in her own unpublished work since the *Cell* paper. But when Diana West of *The Washington Times* asked if Imanishi-Kari meant that she had duplicated the central claim of idiotypic mimicry, the scientist explained that she had found the idiotype on the endogenous gene but didn't know whether idiotypic mimicry was the cause. "That's one possible mechanism. Another possible mechanism is that there is a pure molecular mechanism. Another possibility is translocation," Imanishi-Kari said.

(The *Cell* paper's own summary states, "The expression of endogenous genes mimicking the idiotype of the transgene suggests that a rearranged gene introduced into the germ line can activate powerful cellular regulatory influences." The paper later mentions the possibility of alternative theories, but it returns to idiotypic mimicry as the more likely explanation for the experiments' results.)

The *Detroit News* reporter also asked Imanishi-Kari if Baltimore still supported her.

"I believe that he does," she answered.

"Have you talked with him about this recently?" Gordon pressed.

She seemed to get defensive: "This is not a press conference from David Baltimore. This is my press—"

Gordon interrupted: "Have you spoken with him recently? And he's expressed his support still?"

Singal jumped in and said that while Baltimore still supports Imanishi-Kari, he hasn't been called by Dingell to testify at the upcoming hearing. "She has [been]. And, in fact, what's happened here is the whole focus of this has changed. The focus originally was on the Nobel Prize–winner to attract a great deal of attention and publicity," Singal charged. "That didn't work out so well, so now I think we're in a more dangerous state where they're training their guns on the little guy, the smaller scientist who, unlike a Nobel Prize–winner, cannot command the resources to the support of her position. And I think in the longer run that is a far more dangerous threat to science than going after the more prestigious names."

That day, Baltimore released to the press a statement affirming his support for Imanishi-Kari: "The Secret Service report contains nothing to change my view of Dr. Imanishi-Kari or her research. The report is very unspecific but finds no fault with any of the research that we reported in the *Cell* paper," Baltimore said. "Unless there is information that has been withheld from us, it is apparent that, after years of investigation, a reasonable person must conclude that the *Cell* paper was an honest report of collaborative research in two laboratories."

The oversight subcommittee's hearing the following Monday was short, just an hour and twenty minutes. Imanishi-Kari attended the hearing but did not speak. In Dingell's opening remarks, he said that Imanishi-Kari was not asked to testify about the Secret Service report but about how she created her key notebook. "Instead her attorney submitted a letter claiming to be confused about why the subcommittee today focuses on the so-called unpublished data. First, it is this data submitted by Dr. Imanishi-Kari that was extensively relied on by the NIH panel in their decision not to retract the paper. Second, it is simply wrong to assert that all this data is unpublished."

Dingell provided for the record a March 28, 1988, letter from Imanishi-Kari to Dr. Katherine Bick, the NIH's deputy director for extramural research. The letter indicated that control data from page 107 of the I-1 notebook were published in Table 2. The tapes on that page, Dingell said, were found by the Secret Service not to be authentically dated.

The committee chairman also expressed disappointment that a number of scientists would not publicly voice their criticism of the *Cell* paper "for fear of disapproval by their colleagues. . . . After all, if science is to police itself, it must be expected that scientists will do the things that are necessary to ensure that in fact it does police itself. If science does not choose to, then perhaps there will have to be other mechanisms of ensuring that this is properly done." Dingell did go on to say that there were signs that the current NIH investigation would properly resolve the allegations.

"The standard in science is, and I believe should always remain, a single simple thing: truth. It is only when allegations are minimized, data is not examined, and people do not behave in a straightforward manner that it is necessary to employ forensic methods to settle questions of fact," he said.

Secret Service agents John Hargett and Larry Stewart were first up before the subcommittee. Their reports for 1989 and 1990 were written, but could not be clearly understood without a scorecard linking the page numbers and tapes to actual experiments. However, it was clear that the agents found serious problems about when the notebook records were produced.

Hargett, Stewart, and subcommittee aide Bruce Chafin traveled to MIT on February 21 to gather information about the counters that were actually used at the building where Imanishi-Kari's lab was located and to collect samples of counter tapes from other scientists who used the same machines.

They found that of the four available machines, each of which could be distinguished by its printing characteristics, only two could have been used to produce the tapes in Imanishi-Kari's notebook. Many of the tapes found in her notebook, the investigators contended, were not consistent with samples dated by other researchers around the purported time of her experiments. Many of the questioned tapes, they said, were on pages that the agents had earlier questioned because of altered dates or impressions.

"As a result of the first and second examinations, it is our opinion that a large portion of this notebook is not authentic with respect to time," Hargett said.

Stewart explained that the counter number on the other scientists' tapes advanced by about twelve a day over an extended period of time. However, they found an Imanishi-Kari tape whose counter numbers advanced by 1,353 in a one-day period. "We don't understand anything about the research, and we never have," Stewart said, "However, we do understand trends."

In response to a question from Representative Wyden, Hargett said that taking the two Secret Service reports together, the agents found about forty-four questionable pages in the I-1 notebook. He added that some of the tapes appeared to have been cut and pasted down in order to fabricate an experiment.

"Based on the counter tapes, I would say it is our opinion that at least one-third of that I-1 notebook is not authentic," Hargett said.

The subcommittee turned next to Suzanne Hadley, who by that time was the deputy director of OSI; the new director was Jules Hallum, a former Oregon Health Sciences University virologist. Hadley told the subcommittee that Imanishi-Kari's "cooperation has not been what we would have hoped." Failure to provide the requested index made the investigators' job harder, she said, making it difficult to locate the data behind a particular portion of the *Cell* paper.

In response to Dingell, Hadley said that NIH had estimated the direct costs of the *Cell* paper funded by federal grants to be $82,503. That total involved the pro rata salaries of Imanishi-Kari, Reis, Weaver, and Imanishi-Kari's technician at MIT, who was one of the co-authors, and the cost of supplies and equipment.

Hadley also said that Imanishi-Kari's renewal application for her NIH grant was not funded and that a second supplemental grant, awarded in January 1990 and running through December 1990, was still in place. Tufts had been informed in April that the application would not be funded.

"Does the right hand talk to the left hand at the National Institutes of Health?" Dingell asked. "On the one hand you have an investigation going and on the other you have a grant being made. Is that something that I should find curious?"

Hadley could only explain that in January, the evidence of problems with Imanishi-Kari's work had not been developed enough to stop the supplemental grant.

At the end of the hearing, Wyden noted that the subcommittee staff and the NIH were finding evidence of data fabrication. "My concern, Mr. Chairman, is that there may have been a number of federal criminal statutes that could have been violated by Dr. Imanishi-Kari." The criminal acts in question included possibly submitting false documents and making false statements to NIH; obstructing the NIH and congressional investigations; and lying in her testimony before the oversight panel.

Wyden suggested directing the Secret Service and the congressional staff to share their evidence with the federal prosecutor in Baltimore, who had jurisdiction over NIH-related violations of federal regulations and federal criminal statutes.

Dingell agreed. He referred the case to the U.S. attorney's office as well as to the inspector general's office of the Department of Health and Human Services, which conducts investigations on NIH matters for the federal prosecutor.

As a matter of course, the OSI had already informed the inspector general that criminal violations might have occurred. However, Suzanne Hadley said that up to that point the inspector general's role had been to meet with the OSI investigators, subpoena the evidence that they needed, and act as the official liaison with the Secret Service.

After the Dingell hearing, Imanishi-Kari continued to deny that she was guilty of any scientific misconduct. "I have done nothing wrong, but I am fully prepared to respond to the committee once I know what I'm accused of," she said, according to the May 15 *New York Times*.

■

On October 13, 1990, a Saturday morning, Thereza Imanishi-Kari and her lawyer sat down with the NIH panel of scientists and the OSI investigators to discuss their questions. Singal started off with a statement of concern about the sharing of information with the congressional subcommittee and the U.S. attorney's office and the danger that politics might taint the investigation.

"We believe that by hearing directly from Thereza, you will have an opportunity to see for yourselves that she is an honest person," Singal said, according to a transcript of the meeting. "You will have an opportunity to weigh her credibility yourself, and we believe that that will help you properly assess the facts presented."

He said that Hallum had assured him that OSI would "not be stampeded into conclusions by political pressures. . . . There are very substantial political pressures at play here and it is our hope and expectation that OSI can resist those."

Singal also cautioned the scientists to be wary of the Secret Service evidence and not to give it credence simply because the Secret Service is an investigative arm of the federal government. "I would urge that . . . you subject the Secret Service reports and analyses to the same type of scientific scrutiny which you would give to other scientific endeavors," Singal said. "Frankly, I do not believe that has taken place to date. I think what has happened is that the Secret Service has done what it has been asked to do by the Dingell committee. Its reports have been accepted blindly, without scrutiny, and I would suggest to you that by any objective scientific standard, those reports simply do not pass scientific muster."

Imanishi-Kari told the investigators that although she pulled together all of her loose notes and pages to form the I-1 notebook before meeting with the first NIH panel, she did not change any of the pages. "In fact I had very strict orders from Normand

[Smith, Baltimore's lawyer] not to touch anything, not to write anything, not to touch anything, not to do anything," she said.

The panel members seemed to become frustrated by Imanishi-Kari's inability to explain the dates in her papers. At one point she said: "I never said dates are not important. Of course, dates are important. But dates to me were never real to me. Dates are never real."

Hallum finally said, "You have given us reasons, four or five different reasons for what the dates mean. What good does it do you to date it at all if you don't put down the reason for the date because obviously at this time you can't remember why you chose that particular date to write down."

Imanishi-Kari later in the interview said, "Fortunately, I have very good memories to experiments. I don't have good memory for dates. And because of my good memory for experiments, you can name a cell line, I can tell you the characteristic of the cell line without looking at anything. So also what was done and how it was done. That is something that I do have a very good memory, which is very good from one side, but very bad because then I don't, I don't record keep very good."

She concluded her interview by acknowledging that "to you it may look very messy the way I put my table. And I never thought that there is a certain way you have to put your laboratory data. . . . In my office is a pile this big of things. If somebody wants something I always know where it is in the mess. It is the same thing with my book. It may look to you sloppy, but to me that is how I understand what I am doing."

Except for the flurry of activity surrounding Imanishi-Kari's news conference, the congressional hearing, and the report to the U.S. attorney, little seemed to be going on publicly. The October 1, 1990, issue of the newsletter *Science & Government Report* said that the case and two others were moving "glacially toward resolution," noting that the NIH was still without a director (Wyngaarden had retired in August 1989).

A *Science & Government Report* interview with Dr. James Watson earlier in the year seemed to set the tone for how many scientists were feeling about the controversy. Watson said the Baltimore case had gone on too long. "I keep hoping that finally NIH will issue a report and David will be cleared and the goddamned matter will end," Watson said. "It's partly not the fault of anyone. It's that you've got lawyers into it. Everyone is trying to protect people with due process, so no one will be convicted unfairly. But it also means you can never be cleared quickly."

Some scientists, though, were becoming convinced that there were serious problems with the *Cell* paper, and with how Baltimore had handled Margot O'Toole's original questions. Harvard's John Edsall had not thought much about the case since testifying in 1988 about the difficulties that whistle-blowers face. That changed, however, when he read a mailing he and scientists around the country received from Yale mathematician Serge Lang, contrasting Baltimore's statements and O'Toole's 1989 testimony before the oversight subcommittee. The Lang letter made Edsall think again about the controversy. Edsall would become one of the "Harvard Mafia."

"I always felt she was honest but I wasn't sure she was correct," Edsall says. "Now I was confident in her being right."

Edsall did not get in touch with O'Toole until several months later, when the OSI draft report was leaked, and the scientific community and the public at large learned that the OSI investigators apparently had not been impressed by Imanishi-Kari's explanations or with her lawyer's complaints about the Secret Service.

FIFTEEN

A Case of Fraud

■

Almost two years after the NIH reopened its investigation into the *Cell* paper, the Office of Scientific Integrity reached its initial conclusions. Copies of the preliminary draft went to the co-authors for comment in early March. Despite the supposed confidentiality of the contents, the explosive findings could not be contained; on March 20, 1991, the report was leaked to the major newspapers and scientific journals. Unlike earlier reviewers, the OSI determined that Imanishi-Kari had fabricated data for the June subcloning experiment and the January 1985 cell fusions done for Table 2, and found many other less definitive but troubling problems with the *Cell* paper.

"It remains unclear if these experiments actually were done. . . . It is probable that a substantial portion of the notebook, the major source of data provided to substantiate Weaver et al., was falsified," the draft stated.

Margot O'Toole's actions in pursuing her doubts, however, were described as "heroic." The OSI draft said that she had "suffered for the simple act of raising questions about the accuracy" of a scientific paper. "She deserves the approbation and gratitude of the scientific community for her courage and her dedication to the belief that truth in science matters," the report said.

The OSI report, which relied heavily on O'Toole's critiques for framing and examining issues about the data, said her assistance during the investigation was "invaluable." That helpfulness contrasted starkly with the attitude of Imanishi-Kari, who—as

OSI's chief investigator, Suzanne Hadley, had testified to Dingell's subcommittee the year before—would not prepare an index of her laboratory records. The report did note that Imanishi-Kari eventually sat for a lengthy interview with OSI investigators on Saturday, October 13, 1990.

The key to OSI's report was the combination of new Secret Service examinations of the radiation-counter tapes and an unusual statistical analysis on the counter data. That analysis was prepared by Dr. James Mosimann, a biostatistician who was a member of the OSI working group; Mosimann was the former head of NIH's lab of statistical methodology. The report also relied on the earlier Secret Service findings; on responses from Imanishi-Kari and others to NIH questions; and on related grant applications. Unlike the earlier Secret Service reports, the OSI draft tied challenged pages to the science done and purportedly done. Altered dates and questionable counter tapes no longer appeared meaningless or trivial. The scope of the investigation went beyond the data that had been published for the original paper, to include the subcloning and other data used to correct and bolster the findings after the fact. "Without unpublished data from these notebooks, the paper would not have withstood the scrutiny of earlier investigations," the draft stated.

The principal focus of the investigation was the I-1 notebook, which Imanishi-Kari compiled before turning her records over to the NIH and the subcommittee in July 1988, and the Moema Reis "R-1" notebook, which included what became known as the seventeen pages. But besides these notebooks, the Secret Service rounded up some sixty notebooks from other scientists who worked in Imanishi-Kari's lab or nearby labs in order to compare her counter tapes with others from the same time period and earlier.

The report noted that the primary data had not been secured by a neutral party when they were first challenged, so there was an opportunity for them to be altered. Also, records were not kept of what precisely was seen by the Tufts, MIT, and first NIH reviews. Imanishi-Kari would argue that the NIH had her notebook before the panel brought up the issue of the subcloning, so she could not have forged the June subcloning records in response to the first NIH review. But the OSI said that, according to that panel, on the morning of its second meeting in June 1988 "for the first time, Dr. Imanishi-Kari presented data that did appear to support the table. . . . Some copies of data from the spiral notebook, copies held by NIH and the subcommittee, indicate that the spiral binding of the notebook was intact at the June meeting; in the version supplied in July 1988 . . . the spiral binding was missing."

Some pages had been copied—some of the copied pages were dated—and provided or cited earlier to NIH officials in response to questions. A letter that Imanishi-Kari wrote to Katherine Bick on March 28, 1988, for example, providing copies of thirty-six pages, that included pages 105–107 and 132 in the I-1 notebook, and 27 and 37–40 in the R-1 notebook; several were dated.

The notorious seventeen pages were copied by O'Toole and distributed to several colleagues, establishing at least a fixed end date for them. O'Toole also identified pages now existing as 41, 83–88, 94, and 113 as records that were reviewed at her meeting with the Wortis group; she had written notes on some of them.

The strongest evidence of fabrication came from the June subcloning data, recorded on pages 124–128A of the I-1 notebook. The data had not been published as part of the *Cell* paper, but the OSI noted that they were required by the first NIH review for a correction. It may be recalled that the first NIH scientific panel, disturbed by the weaknesses of the paper, would have been ready to throw it out had Imanishi-Kari not produced the supplemental data.

These data were the subject of O'Toole's formal complaint to the OSI, on November 6, 1989. "The data shown in I-1 pages 124–128A are fabricated," O'Toole wrote. "Other than the experiments done by Dr. Reis on the cells from the Table 2 fusion, no other data were generated from these hybridomas before the paper was published. The gamma counter tapes on these pages will, I am sure, turn out to be from one of the two machines on our floor. However, the tapes will, I am sure, match others from this counter done at another time. This will probably establish that the data are for an unrelated experiment done at a different time, a time when cells from the Table 2 fusion could not possibly have been involved."

Imanishi-Kari, however, told NIH in 1990 that the cloning had started in June 1985 and that pages 124–128A were created in 1985, or at the latest in January 1986. In a response to the first NIH panel in July 1988, she said that subcloning of certain hybridomas were done between June 7 and June 9, 1985.

In a December 30, 1988, letter to NIH, Imanishi-Kari said:

"At no time did I say to Dr. O'Toole that the subcloning analysis of wells in Table 2 was not performed."

In a telephone interview with the OSI, Moema Reis first said, "I don't really know what you mean by the June subcloning," according to the report, although she later referred to the pages and stated: "Yes, I was there. We did this together. I did most of the cloning and Dr. Imanishi-Kari did part of the cloning because it was too much for me to do myself." The OSI report contrasted this statement with Imanishi-Kari's explanation that the purpose of the experiments was to give Reis practice in cell fusions and to isolate transgene-producing clones. Both women said it was Imanishi-Kari who wrote the entries in the records.

As to the Wortis group's contention that in O'Toole's absence it had seen subcloning data, the OSI report said that it was uncertain exactly what documents had been reviewed at the informal meeting in 1986: "The testimony of the members of the Wortis committee suggests they saw something they took to be subcloning data for the wells reported in Table 2 of the *Cell* paper." The OSI also noted that when Dingell asked Wortis when he first learned of the June subcloning data, the Tufts scientist said, "I don't know which June subcloning data you are talking about."

Imanishi-Kari would later complain that "June subcloning" was NIH jargon and not a phrase used around her lab, so it was unreasonable to make anything of Wortis's and Reis's ignorance of the term.

This back-and-forth would prove nothing, however, and the OSI turned to the Secret Service forensic studies and to the statistical analysis of the data.

The crucial pages were dated 6/20 to 6/22, with no designation of the year, and page 124 was headed "Test of subclones from Moema." Most included a single column of green counter tapes and three columns of handwritten counts. Such colored tapes were found in only two sections of the I-1 notebook.

The sketchy May 4, 1990, Secret Service report to Dingell said the tapes on those pages were "not consistent with experiments having been performed by other researchers on or around those dates." The OSI then asked Agents Hargett and Stewart to determine when the tapes were likely produced. The agents searched more than sixty notebooks for tapes of the distinctive green color, including records from Herman Eisen's lab, located on the same floor as Imanishi-Kari's. The agents found numerous samples, including tapes in the records of Charles Maplethorpe, the former graduate student who originally alerted Stewart and Feder to O'Toole's predicament; of Mary White-Scharf, a postdoc in Imanishi-Kari's lab from 1981 to 1985, who used to work with her in the Rajewsky lab in Germany; and of Mark Pasternak, who worked in Eisen's lab in the early to mid-1980s.

John Hargett and Larry Stewart found that all the green tapes in the other notebooks were dated 1981 through January 1984; there were none after January 1984, except for those in Imanishi-Kari's I-1 notebook.

Comparing the tapes for paper color, type font, and ink, the agents declared a "full match" between the June subcloning pages and tapes from Maplethorpe's book, dated 11/26/81 and 4/20/82.

The agents also studied the ink used to produce the gamma tapes on several pages of the I-2 notebook, belonging to Philip Cohen, one of Imanishi-Kari's postdocs. Those pages, which were not questioned, were dated 6/3/85 through 6/28/85, bracketing the June subcloning pages. The agents determined that both sets of tapes were produced with the same counter machine, but Cohen's tapes were of a more common yellow variety and the ink did not match.

The agents had found compelling evidence that the June subcloning data had been fabricated. Simply stated, Dr. Imanishi-Kari could not have done the subcloning in 1981 or 1982—or even harvested the hybridomas for later use—because she did not have the transgenic mice then.

The OSI did not depend solely on the Secret Service findings. The statistical evidence also pointed up problems with the handwritten data on those pages. As would be expected, the gamma counts on Imanishi-Kari's tapes fit a Poisson distribution, an analysis differing from a normal curve and often used in biological experiments where the variable is limited to the number of cells in an area. The handwritten counts in an uncontested section of the R-2 notebook also fit the model. But none of the I-1 handwritten counts for the subcloning data matched a Poisson distribution.

The OSI also developed a "spikiness" index to test the pattern of radiation counts in which adjacent counts are widely disparate. Again, while the spikiness of the tape samples and the R-2 handwritten counts was consistent with Poisson distributions, the handwritten subcloning data were not. Routine rounding was not a factor; the subcloning data "actually reflect anti-rounding behavior," the OSI report said.

By statistical analysis, the OSI determined that the handwritten records, even if the numbers were rounded, did not reflect numbers generated objectively by a machine; the digits were too studied. They must have been made up.

"The behavior of the data is consistent with their having been generated by a conscious or unconscious selection procedure that avoids some numbers and favors others," the draft stated. "In short, it seems impossible that the June subcloning handwritten counts were produced as a result of the experimental procedure described as having generated them. The counts therefore are not authentic, and must be considered to have been fabricated." The OSI report would return to the June subcloning data, and their relationship to the seventeen pages found by O'Toole, in a discussion of some questions about the paper that did not reach the OSI's threshold for fraud, but were nonetheless troubling.

The forensic and statistical analyses of the January 1985 fusion data, however, were strong. These suspect data, on pages 101–107 of the I-1 notebook, were significant to the paper because some of them were published in Table 2 and because, overall, the data supported the *Cell* paper's conclusions, the OSI said.

Page 101 of I-1 briefly described the fusing of cells from the spleen and lymph nodes of four mice—two transgenic, one normal, and one nontransgenic littermate of an engineered mouse—with tumor cells to create hybridomas for study. The other pages record the results of the radioimmunoassays of the hybridomas, including tests for the idiotype. The idiotype results for the nontransgenic mice were used in Table 2.

The results from the transgenic mice were not used for Table 2, but the data on the nontransgenic mice were pooled with the results of another fusion and published.

The tapes, supposedly from a continuous experiment, were in sets on pages 102–107: A green would be on the left side and a yellow would be on the right. The Secret Service found a full match for the green tapes on pages 102, part of 103, 104, 106, and 107 with green tapes dated April 1982 to January 1984—well before the experiments related in the *Cell* paper—from the notebooks of Maplethorpe, Pasternak, and Edward Reilly, another scientist in Eisen's lab.

A portion of green tape on page 103 was found to have been printed in a different ink than that used for the rest of the green tapes. That segment was matched with tapes dated November 1981 through April 1982 from Maplethorpe's notebooks. The page 103 segment also fully matched the green tapes found on the June subcloning pages, records that OSI said were fabricated.

These Secret Service findings indicated that the tapes had been produced substantially earlier than the dates of the purported fusion experiments, suggesting that Imanishi-Kari had used tapes from an unrelated experiment to make up data. If this was so, the OSI investigators thought that statistical analysis might produce some sort of confirmation. They believed that the counts of radioactivity in a fusion experiment probably would show little, if any, pattern.

The OSI team found that the green tapes, however, did follow a cyclic pattern, unlike the tapes from fusion experiments in the Reis and Weaver notebooks. Furthermore, the segment of green tape on page 103 reflected a different cycle, which indicated to the OSI that unrelated data were inserted in the middle of data on pages 102–104.

The OSI also reviewed pages concerning experiments with the Bet-1 antibody reagent, which was reported to bind strongly to the mu^{a} transgene (the co-authors had already corrected their misstatement that the reagent bound only with the transgene). "The Bet-1 reagent was key to the reliability and accuracy of the results reported in the *Cell* paper . . . [and] played a crucial role in the detection of the transgene," the draft said. However, the investigators were mistaken when they wrote that "a failure . . . to react with appropriate specificity would open the possibility that Dr. Imanishi-Kari was in many cases detecting transgene, rather than endogenous genes." In fact, that failure would cause her to pick up more endogenous genes and designate them as transgenes, and thus would weaken the paper's conclusions.

The draft dealt with the records that indicated whether Bet-1, when it was "labeled"—tagged with a radioactive substance such as iodine—worked and was "good," or did not work and was "bad."

Pages 110–119 recorded difficulties with the reagent. The pages were dated March 17–21, 1985, and listed experiments in which the reagent was first bad and then good. An experiment on page 121, dated May 22, 1985, also is good. While these pages reflected problems with Bet-1, they showed that the difficulties "were not insurmountable"; as long as a batch could be made that worked properly with controls, it could then be used further in other experiments to show whether the transgene was present. Without control tests, the scientists would not know whether they could trust the results of the reagents in other experiments.

The records, the OSI contended, showed that Imanishi-Kari's lab knew of difficulties with the reagent before deciding which data were publishable and which should be discarded. According to the draft, Imanishi-Kari said that problems with the reagent were key to determining which data in the seventeen pages were publishable.

On page 18 of R-1 (which included the seventeen pages), dated May 7, 1985, Bet-1 reacted equally well to transgene and endogenous standard proteins. Data from tests using the reagent (on pages 31–34 of R-1, dated May 23–30) were published in Table 2 of the *Cell* paper; retests on June 6, recorded on the same pages, were considered invalid because of problems with Bet-1 controls.

So what the investigators had found in the I-1 notebook were dates on which the Bet-1 purportedly was not functioning properly before and after the dates (May 23–30) for the published Table 2 data. No control tests indicating that the reagent was working at the time of the May 23–30 experiments were found in the seventeen pages.

However, the experiment recorded on page 121 in the I-1 notebook, if valid, would show that the reagent was functioning properly on May 22, and thus could support the data from May 23 to May 30. The data on page 121 were published in the authors' May 1989 correction to show the specificity and sensitivity of Bet-1.

Tracking the pages and dates was a painstaking job that required examining the handwritten notes and special characteristics of the counter tapes on each page.

Some of the pages from 110 to 121, which listed the tests of alternately bad and good iodinated, or radioactive, Bet-1, had Beckman gamma-counter tapes that included not only the data counts but also register numbers that showed how many pages of counts had occurred. The register numbers are recorded by the gamma counter and would not be affected if the counter was hooked up to different printers. The Secret Service found, however, that most of the tapes in the I-1 notebook were trimmed so that the register numbers were eliminated. Some of the tapes also had "position" numbers that identified the order of samples for counting.

Page 110 was handwritten and recorded tests of newly made Bet-1 and AF6 reagents. The bottom half of the page was dated 3/19 and bore the notation "Test on a.17.2.25." Pages 111 and 112 had untrimmed Beckman gamma-counter tapes with consecutive position numbers: 96–116 and 117–135. Page 111 had "Thereza 3/20/85 a.17.2.25" written on the tape. Page 112, with a handwritten note indicating a bad reagent, had a register number of 01447. Page 113 also had a handwritten note indicating bad Bet-1; the tapes on that page were trimmed. The OSI draft said that the radioimmunoassay counts for Bet-1 and AF6 against standard proteins (pages 111–113) also showed that the Bet-1 was not working properly.

Now the OSI tied page 114, dated 3/20/85, to page 110 because of a notation on page 114 that said "label AF6.78 (3/19). Bet-1 label again !! less time and 0.5 mg/ml CT." Apparently the AF6, iodinated on March 19 and recorded on page 110, was to be used in the test documented on page 114.

Position numbers were found on untrimmed tapes on pages 115–18, 153–79, 180–200, 201–218, and 219–33. The tape on page 115 was dated 3/21/85 and had "Thereza" and "Bet-1" written on it. The tape also had radioimmunoassay counts marked off by hand in sets of six and continuing on to page 116. The labels agree with details of dilutions on page 114.

Page 116 has a tape with the register number 02799 and three more sets of radioimmunoassay counts. Tapes on pages 117 and 118 recorded tests on the same materials but with the AF6 reagent; the page 117 tape had a register number of 02800.

The OSI believed that page 119, which was dated 3/22/85 and had trimmed counter tapes, was also "strongly associated" with page 114 because of its listing of radioimmunoassay counts for the two reagents. Page 120, on the other hand, did not appear to have any "clear connection" with the pages before it or with the page following. As previously mentioned, page 121, dated May 22, 1985, referred to the iodination of the two reagents and to "Pool 1"; handwritten data indicated that the Bet-1 worked properly in a control test.

According to the Secret Service analyses, the Beckman gamma

counter tapes on pages 111 and 112, dated March 20, 1985, and those found on pages 115–118, dated March 21, could not have been made one day apart, nor could either of these sets of tapes have been part of an experiment performed on or about those particular dates. The ink intensities were different, but neither reflected a newly changed ribbon. Comparison to tapes in the other notebooks established the fact that it was impossible for tapes produced a day apart—as tapes on pages 113 and 116 purportedly were—to show an advance of 1,352 in counter numbers. In the Maplethorpe and Weaver notebooks, the daily change in counter numbers over a long period of time was roughly 12.

The Secret Service said the purported March 20, 1985, tapes were "most consistent" with tapes of other researchers around January 17, while the March 21 tapes were "most consistent" with tapes for experiments dated around December.

The agents also determined from their analysis of indentations on the paper that the date on page 119 was changed from what appeared to have been 12/16 of some ungiven year to 3/22/85, and that the page was produced on top of a page dated 8/26/84. Their earlier indentation analysis also showed that page 121, dated 5/22/85, was prepared before page 5; page 5 appeared to be associated with pages 3 and 4 and there were August 1984 dates on page 3.

While Imanishi-Kari acknowledged in an interview with OSI investigators that the advance in register numbers was too great for a single day, she had little explanation for it. She cast doubt on the meaning of the dates on the pages and she tried to disassociate page 114 from pages 115–118, as well as page 110 from pages 111–113.

"The dates don't mean anything and the dates definitely can't mean the date that the counter came out," she told the OSI about the dated tapes with the discrepant register numbers on pages 111 and 115. "Just for the simple fact that if they came from the same counter . . . it is an enormous number of page difference." But she did not have any recollection that they came from different counters.

Then why did page 114 appear where it does, obviously between 113 and 115? "Well, I don't know. I think it is because I found [it] in my folder and I just left it there," she said.

Imanishi-Kari also could not confirm the May 22, 1985, date on page 121, which recorded the good Bet-1 test, nor could she say if it was that batch of reagent that was used for the Reis experiments on May 23–30. Yet in her testimony at the 1989 congressional hearing, Imanishi-Kari referred specifically to page 121 and indicated no uncertainty about its May 1985 date.

The scientist's contentions that the dates were meaningless or arbitrary were not credible to the OSI investigators. As they noted, Imanishi-Kari explained in her March 1988 letter to NIH's Katherine Bick that "careful tracking of dates" is necessary because the radioactive reagents deteriorate over time. "The iodinated anti-allotypic antibodies [like Bet-1 and AF6] were always tested on positive and negative controls and used within a short period of time, normally not more than two weeks after iodination," she wrote to Bick.

The OSI draft also pointed out that the dates recorded by Imanishi-Kari were important to her colleague, David Baltimore. During his 1989 testimony, he said that he saw her papers with dates and notations and that she was "not depending on her memory entirely for that." But more important, in his interview with the OSI investigators, Baltimore explained that it was crucial to know when the Bet-1 worked properly on controls: "When Thereza shows me data like this I know that it can work, and so the real question is, in those experiments which we relied upon, was Bet-1 functioning or not functioning, and I have gone back and looked at that and the answer is Bet-1 was functioning in those experiments. So certainly the argument that the data in the paper comes from days when the controls worked is true, and that's what strikes me as important is that the controls worked."

The investigators determined that if the handwritten dates on the I-1 notebook pages were meaningless, then there were no data for the Bet-1 controls for the published data from the R-1 notebook.

"Faced with all these unexplained discrepancies, there remains an obvious explanation," the report said, "namely, that the purported record of experiments on I-1 pages 110–121 is a falsified record prepared to answer the challenge raised by Dr. O'Toole concerning the reliability and specificity of the Bet-1 reagent, and also to demonstrate that data recorded in the 17 pages indicating shortcomings of Bet-1 were offset by other laboratory data that showed such shortcomings were surmountable."

The OSI also examined the other data that were fundamental to the central conclusion of the *Cell* paper—the results of the isotyping experiments which determined whether a cell produced mu (IgM) or gamma (IgG) antibody. A gamma antibody would indicate an endogenous gene. The authors had already said that the isotyping was not done as stated on the Table 2 cells, a fact that OSI investigators did not excuse. They then looked at the Table 3 experiments, which included isotyping. Hadley said that Table 3 was not an acceptable substitute because the hybridomas were not randomly selected. The investigators also found serious weaknesses in the Table 3 data.

The characterization of the hybridomas in Table 3 was also a significant factor in the issue of double producers. These cells produce antibodies with different isotypes, indicating the production of both transgene and endogenous antibodies. With a double producer, the positive idiotype test could be a result of the foreign gene, not the mouse's own gene. The experiments would still have shown some interesting results about the impact of a foreign gene on the immune system, but the presence of double producers would have demolished the argument for idiotypic mimicry, which had made the *Cell* paper so special.

Table 3, a collaboration by Imanishi-Kari and Weaver, was actually done before Table 2. Imanishi-Kari studied the antibody properties and did some DNA and RNA analysis, while Weaver worked on the DNA and RNA characterizations of the cells. Weaver, as well as the other co-authors, had described the collaboration between two approaches—molecular and serological—as an important way of cross-checking the experimental results.

Weaver's DNA and RNA analysis in Table 3 indicated that a majority of the cultures did not show any transgene RNA, and the few that did retain RNA corresponding to the 17.2.25 idiotype seemed to have lost the transgene's DNA. This seemed to be evidence that the idiotype had not been produced directly by the transgene. O'Toole, however, had argued that Weaver's probes had not been sensitive enough to detect low levels of the transgene, and that the hybridomas were actually producing transgene antibodies as well. The Stewart and Feder analysis of the seventeen pages also determined that most of the hybridomas were positive for the transgene.

The OSI found unexplainable discrepancies in the characterizations of the hybridomas. Unexpectedly, the OSI said, the records indicated that over time the hybridomas "acquired rather than lost properties," and references to some of the subclones appeared on pages dated before the subclones could have existed.

Also, some of the hybridomas were described differently in Imanishi-Kari's portion of Herman Eisen's "program project" application for a NIH grant as well as in her own application for an American Cancer Society grant. Some of the important isotype characterizations in the I-1 notebook were on pages found by the Secret Service to have been altered in various ways, including changes in isotype.

The OSI draft said that an example of the discrepant characterizations could be seen with the L3.9.4 hybridoma, reported in the paper as a "gamma 2a." According to data in the I-1 notebook, the characterization of the hybridoma changed from mu+ + + to mu− and at one point as a different type of gamma than published. In grant applications and a letter to the American Cancer Society the hybridoma became a "mu kappa." In an October 1990 interview with the OSI investigators, Imanishi-Kari said the reference in the Eisen application was a clerical error; L4.9.4 was the mu hybridoma, while L3.9.4 was gamma2-b.

The explanation did not satisfy the OSI investigators. "The existing discrepancies in the characterization of hybridomas in purported contemporaneous documents indicate that Dr. Imanishi-Kari failed to fully and accurately report in project-related documents . . . the critical characteristics of hybridomas as known at that time from primary data[,] or [that] she operated with a serious degree of sloppiness in handling the hybridomas, thereby permitting notable errors that affected the development of the research and persisted through several requests for clarification," the draft stated.

"A more serious possibility," the report suggested, "is that Dr. Imanishi-Kari falsified the characterization of hybridomas in the notebooks in an effort to have the notebooks support the claims in the *Cell* paper."

As for actual double producers, the *Cell* paper said the scientists had found none in their experiments. But the OSI found that four of seven wells for which data were recorded on page 44 of the I-1 notebook included double producers, reacting positively for mu and gamma. On page 108, OSI found evidence of twenty-two double producers.

Imanishi-Kari insisted that the combination of serological and molecular data did not show that the engineered mice had double producers. Where her serological data might have indicated the presence of double producers, she said, the scientists decided that they actually had "double clones," wells with two cells producing antibodies with different isotypes. "We discussed often the issue of hybrid molecules. . . . We did not discuss the 'double producer' issue specifically because we did not have any experimental results that suggested that those 'clones' were double producers," she wrote the OSI in 1990.

The investigators also expressed concern about the nature of the collaboration between the two laboratories, noting that in two instances where the serological and molecular analyses differed, Imanishi-Kari disregarded the serological results. The *Cell* paper, however, implied that the serological and molecular data corresponded. The L3.9.4 hybridoma was one such case. Another was L4.13.2, the controversial hybridoma in David Weaver's published Figure 4 photograph that showed only a gamma isotype. The serological results showed positive results for both mu and gamma.

Imanishi-Kari told the OSI that serology was only one of the tools she used to characterize antibodies. "It is my procedure, generally, not to use serological results alone. Therefore, in order to define whether a cell expresses a certain determinant or not it has been my approach to use the combination of both techniques: serological and biochemical. To interpret the serological results obtained from the analysis of transgenic hybridomas would have been impossible without molecular analysis."

Weaver told the OSI that he "never felt there was a moment" when any of Imanishi-Kari's work "wasn't available to me to be looked at," but he did not recall discussing the existence of double producers, nor did he believe they were "really relevant to the story," according to the OSI report. Weaver argued that it was clear that he was finding positive idiotypes without transgene RNA. The presence of a double producer would have been "worrisome," he said, but it would not have detracted from a view of the entire population of hybridomas.

The OSI also quoted Weaver as stating that he was unaware of any cases in which a hybridoma isotype's characterization changed over time in the lab records. "I do not recall any conversations or reviews where significant double expressors were discussed. . . . I was not aware that Dr. Imanishi-Kari considered my molecular studies definitive. My molecular studies were conducted early in the characterization of this set of hybridomas. I don't remember any inconsistent characterization of clones, between isotyping and molecular analysis."

The report suggested that had Weaver been "fully aware of the significance of 'double producers' " he might have designed his experiments differently to ensure that he would be able to detect the transgene and he might have been more sensitive to the presence of the light mu smudge in Figure 4.

The OSI examined the Figure 4 photograph controversy, noting that Weaver and Baltimore argued that the important element to keep in mind was that the photograph showed that, according to the Northern blot experiment, most of the hybridomas did not produce the transgene. However, the report said that it was misleading for the authors not to state in the paper that the photograph was a composite based on differing exposures and that a band consistent with mu had been detected, because reviewers might have judged the paper differently had they known. The report said that Weaver's contention that he was not fully informed of the inconsistent isotyping of the hybridomas "provides an explanation for his failure to explore the significance of the occurrence of double expressors."

The OSI investigative team and the NIH panel believed that "as first author, he had a responsibility to disclose" the mu band, as did Baltimore. They noted that Baltimore was Weaver's mentor and a co-author of the paper, and that he testified about Figure 4 at the Dingell subcommittee hearing.

The OSI also raised questions about the authenticity of pages 7–10 in the I-1 notebook, which included data documenting the specificity of guinea pig and rabbit anti-idiotype reagents. (These reagents, obviously, were central to the research on a paper on idiotypes because they would be used to test whether the transgene idiotype was present.) The pages recorded handwritten counts, purportedly from a gamma counter. Despite the investigators' suspicions, however, the report noted that the evidence that these pages were fabricated was not as clear as in other instances.

According to the Secret Service, the 8/22/84 date on page 7 had been altered, although the agents could not determine the original date. The agents also found that pages 7, 8, and 9, dated 8/26/84, were overlaid by page 119, dated 3/22/85, while all the sheets were still attached to the same notepad. This indicated that the 1985 page was written before the purported 1984 pages. Page

119 was already suspect: The agents had told the subcommittee in 1990 that the date on the page had been changed from what appeared to have been 12/16 to 3/22/85 and the tapes on that page could not be associated with tapes from other scientists around 12/16 "of any provided year." The page 119 data showed that the Bet-1 and AF6 reagents behaved appropriately with the culture fluid surrounding hybridomas and with standard proteins, thereby showing that idiotype-positive hybridomas did not express the transgene.

Statistical analysis of the handwritten data on pages 7–10 also showed that the radiation counts did fit the expected distributions, unless unusual rounding had been done.

The report did not find clear evidence of fabrication of data involved in the paper's Figure 1, for which some of the data were missing from the Moema Reis notebook. It was partly O'Toole's discovery that the data were missing that led NIH to reopen its inquiry in 1989. Reis explained that she had done the work for the fourth point on the figure the day after the first three and had recorded the data directly onto the graph.

"These explanations indicate that, at best, there was sloppiness and laxness in recording of data," the report said. "However, there is no clear evidence pointing to fabrication or falsification of the data."

Even if the published results were correct, the OSI noted, they would be meaningless if the actual research was fraudulent. Nonetheless, the OSI also examined scientific papers by other scientists, which had been put forth by Baltimore and Imanishi-Kari and their supporters as evidence that the *Cell* paper's science had held up. OSI said that the papers, indeed, supported some of the paper's conclusions, but that they did not confirm—or disprove—the central conclusion concerning idiotypic mimicry.

■

In addition to evidence of outright fabrication, the OSI investigators said, they found information in the I-1 notebook and the seventeen pages that cast doubt on whether the *Cell* paper was "an accurate reflection of the serologic results." Specifically, the report stated:

1. Expression of the transgene was much more common than reported and could have explained many of the findings of idiotype-positive endogenous antibodies.

2. The Bet-1 reagent, as it was used in the Imanishi-Kari lab, was so poor at discriminating between the endogenous M type of immunoglobulin and the transgene that the allotyping was unreliable.

3. The tests did not show clearly whether the idiotype was present on molecules that did not have the constant region of the transgene.

The report noted that the I-1 notebook included two types of experiments that strongly supported the central serologic findings of the paper. One was an enzyme-linked immunosorbent assay, or ELISA, which is more sensitive than the radioimmunoassay also used for detecting the idiotype, and the other was a set of antibody isolation studies on pages 40–43. These experiments seemed to prove that many of the March 1984 fusions for the Table 3 hybridomas produced endogenous heavy-chain constant regions along with the 17.2.25 idiotype, but the transgene was not detected.

One of O'Toole's contentions had been that the ELISAs had not been done as described, on a plate first coated with anti-idiotype reagent to ensure that the antibodies being tested were idiotype-positive. Rather, O'Toole said, the ELISAs were done on plates coated with anti-mouse immunoglobulin; this could have increased the amount of data showing supernatants that tested positive for the isotype without actually proving the presence of the idiotype. O'Toole claimed that Imanishi-Kari had told her that the test was done only for isotype, not for idiotype. The ELISA contrasted sharply with the radioimmunoassay, but the raw data couldn't be compared because the radioimmunoassay records were now missing.

Imanishi-Kari told OSI that she never tested for isotype with the relatively nonspecific anti-mouse antibody coating.

If the ELISA had been done as reported, a test result would be positive only if there were molecules that were both idiotype-positive and had the specified isotype. Plates coated only with general anti-mouse Ig antibody would show a variety of isotypes present from the ELISA, but would not indicate the presence of the idiotype.

There was little forensic evidence on this issue. A notation, for example, at the top of page 83 indicating that an "anti-17.2.25 coat" was used was added in a different ink and at a different time from the rest of the page, according to the Secret Service.

Despite this lack of substantial forensic evidence, the OSI report said there were other inconsistencies that cast doubt on the ELISA's validity. The report noted that the results from the ELISA varied sharply with the results of a radioimmunoassay done on the same hybridomas. These results were reported in a footnote to Table 3 submitted by Imanishi-Kari as part of her section of Eisen's grant application. While the ELISA showed that most of the samples were positive for the idiotype, less than half were positive based on the radioimmunoassay reported in the grant application.

It seemed unusual to the OSI that the raw radioimmunoassay data, important enough to be reported in a grant application, would be missing from the lab records. The report further suggested that such a great variance between the two tests would indicate that one was unreliable. Yet no one could compare the raw data from the two assays because the radioimmunoassay data were missing.

"Very often, at that time, when I made a pile of data, I threw the original data away," Imanishi-Kari told *Science* in an interview shortly before the report was leaked. She also told *Science* that she used the radioimmunoassay results in the grant application, rather than the more impressive ELISA results, because the transgenic project was "a minor part" of the application and radioimmunoassays were used for other data in it.

The OSI investigators were disturbed because if the ELISA was valid as reported, it was most likely showing many false positives or providing evidence of many double producers; either outcome would seriously detract from the central conclusion of the paper. "It appears that the choice is between believing the serologic methods employed were seriously inadequate or the ELISA was mislabeled and misrepresented," the report stated.

Although Imanishi-Kari argued that there was no reason to make up the June subcloning data, since they were not used in the original paper, the OSI suggested that the data may have reflected a response to the questions Walter Stewart and Ned Feder raised in their manuscript, and were therefore an attempt to "supplement flawed data used in Table 2." Indeed, as has been mentioned, the first NIH review panel was ready to dismiss the *Cell* paper because of its weaknesses, when Imanishi-Kari brought forth the subcloning records.

The most direct connection between the seventeen pages and the June subcloning data, OSI said, was the second part (pages

25–34) of the R-1 notebook. These pages recorded cell fusions done May 8, 1985, and tests done on the supernatants from wells, dated May 23–30. The fusions were to produce hybridomas from the spleen and lymph node of a normal and a transgenic mouse. The data from the control mouse allegedly were not used because the mouse had actually been mistyped and was transgenic—its cells produced many idiotype-positive antibodies; the results from the other mouse were used in Table 2. Some wells were retested June 6, and the June subcloning data were reportedly derived from them.

The OSI said that using the data from the seventeen pages "reflected poor scientific judgment," because (as Stewart and Feder had noted), an arbitrary cutoff had been used to distinguish positive and negative results for idiotype. Also the assay with the AF6 reagent, testing for endogenous antibodies, was "dubious" because eight of the ten wells recorded as positive in the lymph-node fusion were next to each other.

The retest of idiotype and Bet-1 on June 6 gave quite different results. On page 34 of R-1, and on part of the seventeen pages, the control data, reportedly for the June 6 retest, showed poor binding results for both the Bet-1 and AF6 reagents, and the retest was considered invalid.

The OSI contended that one scenario for explaining why the June subcloning data had been fabricated was that when Imanishi-Kari's poor data from the seventeen pages "could not stand up to scrutiny," she made up "unusually clean and convincing subcloning data from these wells as a way of legitimizing the use of the suspect data." Either the subcloning was never done, or it was done but had given "unconvincing results."

The first half of the seventeen pages was not directly related to the subcloning issue, but the OSI noted that controls reported in that section showed a lack of specificity of Bet-1 and low reactivity by AF6. This was another instance, the OSI said, that revealed problems with the Bet-1 reagent; it also pointed to "less discussed problems" with the AF6 reagent.

The OSI charged in its draft report that Imanishi-Kari "repeatedly presented false and misleading information" to NIH and its scientific review panels. Her statements about the June subcloning data and the January fusion experiments "were known by her to be false, or were provided by her with reckless disregard for the truth."

Her statement that various dates on the lab records were "meaningless," the report said, "does not answer the challenges to the paper, and in fact, throws into question the entire fabric of the purported experimental record." The OSI also did not accept Imanishi-Kari's argument that unpublished data were not germane. The investigators threw her own words back at her: "At her October 13, 1990, OSI interview she stated: 'Everything in my books has relevance, everything. The bad data and the good data, everything is relevant for us to know what is going on.' The OSI believed this assertion is unqualifiedly true."

Hugh McDevitt and Ursula Storb, two of the original members of the first NIH panel, submitted a minority opinion, which disputed such weaker findings as those on Bet-1, the double producers, the discrepant characterizations of the hybridomas, and the Figure 4 photograph.

"While the inconsistencies which are cited in these sections are certainly disturbing, and are compatible with the conclusions given in the report, alternate scenarios are sufficiently plausible that we cannot agree with the conclusions as stated in the report," McDevitt and Storb wrote. They did not detail those alternate scenarios.

McDevitt and Storb were also dubious about the statistical analyses used to bolster the findings of fabrication. While they found the analyses interesting, they said the tests were "new and untried" and "certainly not adequate grounds for an allegation or conclusion of fabrication."

The Secret Service forensic tests were "much more convincing," although the two scientists believed that they would need to review the actual chromatograms (filter papers) of the ink samples to determine if the Secret Service findings truly were reasonable. Nonetheless, McDevitt and Storb basically agreed that the June subcloning data and January fusions were not what they seemed.

The forensic results and other inconsistencies on the pages "make it seem likely that the data on these pages are not the result of experiments performed at or near the time stated, but in fact, are data from other experiments performed as much as three years earlier," said McDevitt and Storb, to whom the subcloning results had seemed critical in 1988. That said, however, McDevitt and Storb still disassociated themselves from the OSI criticism of Baltimore and Weaver.

■

David Weaver, who had worked under Baltimore, also came in for some OSI criticism, but of the slap-on-the-wrist type. He was scolded for not appreciating the importance of the double producers and for not noting the various exposures of the Figure 4 X rays and the presence of the mu band. The OSI decided there was no reason to conduct any further inquiry into Weaver's actions.

No inquiry into Baltimore's actions was recommended either, but while the OSI did not charge that Baltimore had been guilty of making up data, its criticism of his behavior since the first Eisen meeting was harsh, considering his exalted position in the scientific community. Most damning were Baltimore's own statements to investigators. The draft quoted him as saying that made-up data in scientific notebooks can't be considered fraud if they are not published, and that if there were fabricated data, then NIH was to blame for pushing Imanishi-Kari into publishing them.

The OSI found Baltimore's statements to be "extraordinary" and "all the more startling when one considers that Dr. Baltimore, by virtue of his seniority and standing, might have been instrumental in effecting a resolution of the concerns about the *Cell* paper early on, possibly before Dr. Imanishi-Kari fabricated some of the data later found to be fraudulent."

Baltimore was let off the hook for his initial strong defense of his collaborator, but the report stated: "It is difficult to comprehend his maintaining this stance as the evidence mounted that serious problems existed with the serological data in the *Cell* paper."

Even though the OSI report was not final—it might be substantially changed after the scientists responded to it—the draft stunned the scientific community. The extensive report seemed to overwhelm even Baltimore, who over the years had staunchly fought attacks on both Imanishi-Kari and the paper itself. But hours after the draft report was leaked to news publications, Baltimore retreated. In brief statements to reporters, he admitted that the OSI draft "raises very serious questions about serological data in the paper. . . . It is up to Thereza Imanishi-Kari to resolve" those questions. He said he would ask his co-authors to retract the paper.

On May 17, 1991, *Cell* published the following on behalf of

Baltimore, Weaver, Costantini, and Christopher Albanese, Imanishi-Kari's technician: "The undersigned four authors wish to retract the article by Weaver et al. . . . because of questions raised about the validity of certain data in the paper. Two authors (Thereza Imanishi-Kari and Moema H. Reis) do not believe that the questions raised have merit and are not parties to this retraction."

Not surprisingly, considering the history of this controversy, Baltimore's retraction would not be the last word.

SIXTEEN

The Duel in Nature

■

The response to the draft of the OSI report in both the popular and scientific press was swift and pointed. Baltimore was no longer the grand Nobel laureate, and Dingell no longer the evil inquisitor.

Coverage was particularly extensive in New York City, the home of Rockefeller University, where many senior faculty members were already questioning Baltimore's ability to lead the research institution. *The New York Times* carried eight pieces referring to the case during March 1991 alone.

The day after the OSI report was leaked, the *Times* ran a front-page article by Philip J. Hilts, headlined "Crucial Research Data in Report Biologist Signed Are Held Fake/Nobelist to Ask Retraction of Paper He Defended." The *Times* noted that the draft report "drew a picture of data falsification that appeared to extend over several years"; the article went on to say that Baltimore had been named president of Rockefeller University despite some faculty members' contention that his handling of the matter "reflected poorly on his ability to lead the university."

On March 22, Hilts profiled Margot O'Toole and quoted both Harvard's Mark Ptashne and Walter Gilbert as critical of the scientific community's response to her challenge. "One of the most surprising things to me is the way so many members of the scientific community and the scientific press were ready to denigrate Dr. O'Toole," Ptashne said. Gilbert, a Nobel laureate himself, said that the scientists in charge at MIT and Tufts failed in their responsibility to a whistle-blower. "The people in authority . . .

failed to investigate properly. Neither . . . [MIT nor Tufts] seriously entertained the question of whether there had been fraud and what should be done."

A March 26, 1991, *Times* editorial compared the case to Watergate and said that while the final verdict might destroy Imanishi-Kari's career, "the most damning indictment should be lodged against the scientific community's weak-kneed mechanisms for investigating fraud."

Wrote the *Times*: "Faced with stonewalling by Dr. Baltimore, one of the nation's most prominent scientists, several investigative panels seemed more intent on smothering bad publicity than digging out the truth. . . . [Baltimore's] decision to tough out the allegations of wrongdoing instead of meeting them openly has ultimately harmed the reputation of the profession he so brilliantly represents."

Nature editor John Maddox contributed an op-ed piece to the March 31 *New York Times,* entitled "Dr. Baltimore's Experiment in Hubris." He had known Baltimore since 1962, yet now Maddox was at a loss to explain his friend's actions. He concluded that Baltimore's arrogance had caused grave damage to the scientific community's credibility: "This case will seem proof that the scientific community can cover up the errors of eminent insiders at the expense of unestablished whistle blowers. The disputed article would not have survived had not Dr. Baltimore been its champion."

A news analysis on the same day noted that while the report's findings were embarrassing to scientists who had opposed the federal investigations, some still argued that the probes had caused unwarranted damage to the scientific community. "I don't want to be quoted as saying that sloppiness or fraud is O.K., but it might be better to tolerate a low level of that rather than create an inhibitory atmosphere in science," said Bernard Davis, the retired Harvard Medical School microbiologist who had staunchly defended Baltimore throughout the controversy.

Scientists at Rockefeller were disturbed over the continuing bad publicity; even though none of the work for the *Cell* paper had been done at Rockefeller, the articles all mentioned that Baltimore was now the president of the university. On March 21, for example, *The Washington Post* ran an article headlined "Scientist Retracts Paper Amid Allegations of Fraud." The first paragraph noted that Baltimore was president of Rockefeller. Such items were greeted by the school's faculty with dismay. "I hope people realize that this had nothing to do with Rockefeller," Dr. Bruce McEwen, dean of graduate studies, was quoted as saying defensively in an April 1 *New York Times* article. "The incident took place somewhere else, in the past, and I hope we won't be victims of the fallout."

John Dingell also appeared prominently in the press. And no longer was he being called the Torquemada of scientists. In an April 21 *Times* article, "Dingell vs Academia/A Crusader Tilts at the Ivory Towers Looking for Old-Fashioned Corruption," reporter Richard L. Berke wrote: "And after years of being attacked by scientists for conducting a 'witch hunt,' Mr. Dingell appears to have been vindicated."

Even sweeter for Dingell, however, was the March 25 editorial in his hometown paper, *The Detroit News,* which took its lumps and trumpeted "Dingell: He Was Right."

The Boston Globe also weighed in immediately with a news story recounting the controversy and a March 25 editorial that criticized Baltimore for his behavior toward O'Toole. London's *The Economist* headlined its March 30 article: "The Baltimore affair / Ignoble." Articles were also run in *Time, People, U.S. News & World Report,* and other publications. A *Time* essay by social critic Barbara Ehrenreich, titled "Science, Lies and the Ultimate Truth," was illustrated with a picture of a man with a test tube for a Pinnochio nose.

The only publication unrelenting in its criticism of Dingell was *The Wall Street Journal,* whose editorial page was the congressman's most ardent attacker. The March 29 editorial, "Politics and Science," stated that "federal investigators now say they've found 'serious misconduct.' . . . What the rest of us have learned is that when politics and science clash, science loses."

In the midst of the reporting and editorializing about the OSI report, articles in *The Boston Globe* and *The New York Times* revealed that MIT and the Whitehead Institute would repay thousands of dollars in federal dollars that they had spent on high-powered Washington lawyers in connection with the Baltimore case. The MIT money amounted to $27,300 in expenses billed by the D.C. law office of Kirkpatrick & Lockhart to prepare MIT officials to testify before Dingell's subcommittee.

Whitehead officials said that it was appropriate for the institute to pay a portion of the approximately $196,000 spent in 1988 and

1989 for legal fees in the defense of the *Cell* paper; the institute eventually repaid about $69,000 to the federal government. According to the April 17 *Globe,* Whitehead paid $100,000 to Akin, Gump in 1988; $12,203 to the Boston firm of Baltimore's personal lawyer, Normand Smith; and $1,776 to the D.C. firm of Galland, Kharasch, Morse & Garfinkle, whose Marc Ginsburg helped early on.

The debate over Baltimore's actions and Dingell's role intensified within the scientific community. Still, no more than a few scientists had actually reviewed the *Cell* paper for themselves. Serge Lang of Yale continued to mail reports on the case, including excerpts from the OSI draft, to scientists around the country.

Harvard's John Edsall, who had been so impressed with Lang's 1990 mailing that he began to speak up on O'Toole's behalf, had not seen O'Toole from the time of the 1988 hearing until after the OSI report was leaked. The eighty-nine-year-old scientist then took her out to lunch to discuss the case further. Walter Gilbert, another O'Toole defender, also had never met her, and would not do so until later in the year. Ptashne introduced O'Toole to Harvard's John Cairns and Paul Doty; it was the first time O'Toole felt she was walking into a room of scientists who believed her.

On April 22, Brandeis University held a debate between Edsall and Bernard Davis. The debate featured impassioned speeches from the floor from Imanishi-Kari supporters as well as critics of Baltimore. Davis "questioned whether the intense public scrutiny of an eminent scientist like Baltimore is 'good for society. . . . In science, some people are a lot more important than others.' When asked if this meant David Baltimore was more important than Margot O'Toole, Davis said, 'Not as a human being. But some people can make more important contributions and the loss of these contributions is large.' " (Davis later contended that he never suggested that Baltimore should be immune from criticism.)

■

Meanwhile, the impact of the OSI report continued to reverberate through Rockefeller University. Norton Zinder had been a supporter of Baltimore's from the beginning, but was distressed when he read the OSI report and its account of the MIT and Tufts investigations. Two years after the 1989 Dingell hearing, Zinder obtained copies of the universities' reports to read them for himself. His conclusion was that Baltimore and the two university reviews had handled things completely wrong. Zinder took it upon himself to visit Baltimore and urge him, for the good of science, to try to be more contrite.

But even before that visit, Baltimore was finally beginning to realize that he had to be at least minimally conciliatory. The press reports were embarrassing to him personally and distressing to the faculty at Rockefeller. Even though David Rockefeller and Richard Furlaud maintained their support of him, his position grew shakier with each passing day. Baltimore read a draft of one possible response to the senior faculty and asked for their comments. He also hired a new and high-powered Washington lawyer: Benjamin Civiletti. The former attorney general (in the Carter administration) immediately went to work trying to find a way for Baltimore to be apologetic and put the controversy to rest.

In early May of 1991, Baltimore issued his anxiously awaited response to the OSI draft report. Still argumentative, he disputed some of the report's conclusions, and claimed that some of his comments quoted by OSI, such as remarks that if Imanishi-Kari had made up any of her data it was because she had been driven to do so by the NIH investigation, did not represent his view.

But the thrust of the statement was Baltimore's unequivocal apology to Margot O'Toole. "I commend Dr. O'Toole for her courage and her determination, and I regret and apologize to her for my failure to act vigorously enough in my investigation," Baltimore said. As for O'Toole's nemesis, Baltimore said he had no knowledge of any misconduct allegedly committed by Imanishi-Kari.

Baltimore insisted that the report did "not undermine" the integrity of Weaver's work. He did not note, however, that the OSI had, in fact, criticized Weaver's work.

With respect to the OSI's criticism of his defense of Imanishi-Kari, Baltimore said his support was "fueled by my respect for Imanishi-Kari's demonstrated abilities as a scientist, by my belief that the paper's scientific conclusions were sound, and by my trust in the efficacy of the peer review process," not by "any lack of regard" for O'Toole. He said O'Toole's "analyses were insightful, her expressions of concern were proper and appropriate, and her motives were pure."

Baltimore noted that the Tufts and MIT reviews had not found any fraud and that the "MIT expert" (presumably Eisen) determined that O'Toole "had correctly identified a minor error."

"I fully expected that this paper, like all others, would be subjected to the rigors of the scientific peer review process, and that efforts by other laboratories to replicate or extend our findings would ultimately test whether they were correct," Baltimore said. But he now realized that he had failed to heed earlier "warnings" and should have avoided public comment until he had the opportunity to study the first NIH panel and Secret Service findings.

"In good conscience, I feared a rush to judgment and I accorded my colleague the benefit of every doubt," Baltimore said. "I now recognize that I was too willing to accept Imanishi-Kari's explanations, and to excuse discrepancies as mere sloppiness. Further, I did too little to see an independent verification of her data and conclusions."

Baltimore added that he "focused narrowly on the questions of whether the paper could stand. . . . In other words, as a scientist, my concern was always for the science: Is the result correct? Can it be replicated and built upon?"

The personal anguish Baltimore had suffered, as well as his own strongly held set of principles, came through strongly. If his statement was an admission of error, it was also a plea to win back his once-unblemished reputation.

Saying he was "shocked and saddened" by OSI's revelations that data had been altered and fabricated, Baltimore insisted that he would never condone falsification by a scientist. "Fraud in the laboratory is not only wrong from a moral and legal standpoint . . ." he said; "it impedes the progress of science, as it makes the review and retesting of hypotheses and conclusions impossible."

He also held out an olive branch to John Dingell. Incredibly, Baltimore, who had so passionately fought what he saw as governmental intrusion into the scientific process, now proclaimed his commitment "to the importance of governmental oversight of federally funded projects.

"I look forward to continuing to participate in a healthy and necessary dialogue to improve the process," he added. Those seemed more like lawyers' thoughts than his own. Baltimore also suggested that better self-policing and recordkeeping "may obviate the need for the repeated hearings and investigations that were needed in this case."

He said that the case showed the need for "clear procedures which guarantee the prompt and thorough investigations of allegations. . . . Questions raised, whether by junior or senior scientists, must be pursued with vigor, and because junior colleagues may be reticent about alleging outright misconduct, it is incumbent upon those more senior to press for a full airing of their suspicions. . . . Scientists must ensure that they do not wait too long or set the threshold too high before calling for the application of close scrutiny to ferret out potential falsity." Baltimore also called for protection for scientists who raise concerns.

Concluding with an apology to O'Toole, Baltimore said, "I recognize that I may well have been blinded to the full implications of the mounting evidence by an excess of trust, and I have learned from this experience that one must temper trust with a healthy dose of skepticism. This entire episode has reminded me of the importance of humility in the face of scientific data."

Many of Baltimore's supporters appeared to be satisfied with his response; the Rockefeller trustees continued to stick by him. But the fissures in the scientific community were becoming more pronounced.

One of the first to criticize Baltimore publicly was Walter Gilbert. He found "very little admission" in the response and was disappointed by its tone. "It reminded me of that moment in the movie *Casablanca,* where Claude Rains stands in the bar and says, 'There is gambling going on here? I'm shocked! I'm shocked!' " he told *The New York Times.*

Gilbert later told *Science* magazine: "The thing that's profoundly wrong with David's statement is that there has been a problem in this [paper] all the way through. . . . By early 1989, when he attacked the committee, it was clear that there was a fraud involved. He'd been told that, and even if he did not believe it, there was still a prima facie case to be answered.

"And yet he chose at that time to do an all-out attack on the committee. . . . He simply refused to notice what was happening—and that's the best interpretation you can put on it."

■

In a prepared statement, O'Toole, stubborn as ever, said she appreciated Baltimore's words of praise, "but his apology does not go to the heart of the question." She repeated her contention that she and Imanishi-Kari told Baltimore back in June of 1986 that some of the experiments had not been performed and that others had not yielded the results published in *Cell.*

Imanishi-Kari, who continued to maintain her innocence, was furious with Baltimore's decision to retract the *Cell* paper. She did not believe that even a correction was necessary. But, at least in hindsight, she was philosophical about his decision to throw her overboard. Rather than feeling betrayed, she believed that the higher up one goes, the more vulnerable one is. She saw Baltimore's vulnerability, his position as a Nobel Prize–winner; he had to try to save himself first. She was hardly naïve.

Imanishi-Kari later told me: "You have to put yourself in David's shoes. [People are saying] 'Thereza's a crook,' and he has to decide between me and doing something for society. I was just the person to use to pull Baltimore down. I don't think it was Baltimore's fault, but the fault of the politicians and the fault of the media."

There wasn't much she could do about Baltimore's defection; however, Imanishi-Kari continued work on her own response to the OSI report. Like Baltimore, she hoped the matter would finally come to rest. Both were encouraged when *Nature* headlined its editorial about Baltimore's statement, "The end of the Baltimore saga."

"To make an error may reflect on a person's judgment," *Nature*'s editors wrote. "But to confess it in the circumstances in which Baltimore now finds himself is a mark of courage. He deserves a break."

■

If there was any break, it was extremely brief. A week after declaring the end of the saga, *Nature* noted at the beginning of another article "The Baltimore saga continues." O'Toole and others had more to say. And Baltimore would be unable to refrain from responding.

The article was O'Toole's response to the OSI report. She blasted Baltimore, his collaborators, and the MIT and Tufts reviewers. While she appreciated the report's praise for her, O'Toole downplayed the heroine's role, saying her "actions were nothing more than should be expected of any scientist." She retold the story of her role and stressed that she herself had at first been willing to give Thereza the benefit of the doubt:

"Failure to replicate a report does not demonstrate that fraud has been perpetrated. Therefore, I did not assume fraud when I observed that the idiotype expression and transgene expression always seemed to go together. It was not until I found the data for the actual published experiment that I knew that the result had never been obtained, and even then I accepted the 'error not fraud' possibility, remote though it seemed."

O'Toole defiantly contradicted Baltimore's claim that until the OSI report there was little proof to indicate that he should consider dropping his defense of Imanishi-Kari. She contended that she showed Baltimore a copy of the seventeen pages at the 1986 meeting with Eisen, Imanishi-Kari, and Weaver, and that he said then that the published results could not be based on those pages. Imanishi-Kari, O'Toole repeated, "admitted that a large series of the published experiments had not even been performed, and that some that had been performed had not yielded the claimed results. . . .

"Since I had made no charge of fraud, a retraction at this time could have been made without disastrous consequences. But Dr. Baltimore said he would oppose a retraction, adding that these kinds of inaccuracies were 'not unusual.' He told me that he would personally oppose any effort I made to get the paper corrected."

When she saw the preliminary report of the first NIH review, O'Toole wrote, she informed the NIH that Imanishi-Kari had told her that the experiments had not been performed, and that that statement had been relayed to Baltimore. "And still he submitted the fabricated data as part of the next correction. He consistently and falsely maintained that my objections were no more than alternative interpretations of valid data. He defended Dr. Imanishi-Kari as 'the most intensely honest person' he knew," O'Toole wrote indignantly. "I consider it a disgrace that Drs. Weaver and Baltimore did not immediately retract the paper in 1986 when they learned of the discrepancies between what was actually observed and what was published."

O'Toole let it be known that it was she, not Baltimore, who was the injured party in this fracas:

"I was left unemployed, and subjected to five years of slander and libel from Drs. Baltimore, Eisen, and Imanishi-Kari," she declared. (And she corrected the OSI statement that her actions had cost her her job, saying that it was "more accurate to say that I lost my career.")

"Because of the coverup at MIT, Drs. Huber and Wortis were able to continue with their slander and libel of me," O'Toole wrote. "I bring this up to document what can happen to a scientist merely for adhering to the professed standards of the profession."

■

Needless to say, Baltimore felt compelled to reply to O'Toole's bitter criticism. He wrote wearily in the May 30 issue of *Nature* that the issues raised by O'Toole had been answered previously. Curiously, while he condemned O'Toole's allegations as familiar, he described them in another paragraph as "new charges, overstatements, and errors."

At the center of his new rebuttal was O'Toole's assertion that the discrepancy was "evident."

Baltimore said that the famous seventeen pages did not obviously prove the experiment was fraudulent. The major element questioned, he said, was data that purportedly came from a control mouse and differed from the published control data. These were results that indicated a high expression of idiotype-positive antibodies, an unlikely result for a normal mouse that did not have the transgene. However, Baltimore contended, as Imanishi-Kari had, that the mouse had been mistyped and that new control data had been generated and published in the paper.

"This issue has been definitively resolved and therefore, Dr. O'Toole's statement that the discrepancy 'remains evident' is incorrect, and her reference to the data she discovered and Dr. Eisen's reaction to it is misleading," Baltimore said. He did not mention the fact that the OSI report asserted the apparent falsification of a page of control data that was purportedly used instead of the allegedly mistyped normal mouse.

Baltimore also disputed O'Toole's charges that he knew that the June subcloning data were false and that he had heard Imanishi-Kari say that she had not performed any of the experiments published in the *Cell* paper or in the subsequent corrections.

His 1987 call for an NIH review of the matter indicated that he believed the co-authors had nothing to hide. Baltimore added that the first NIH panel had reviewed O'Toole's response to its report and decided against modifying its report, making his acceptance of the data as authentic justified.

The initial Secret Service findings of date alterations "seemed consistent with [Imanishi-Kari's] reorganization of her material," Baltimore said, also contending that he had never made any secret of the fact that Imanishi-Kari's laboratory records had been a mix of notebooks, loose pieces of paper, and data sheets.

Baltimore also took issue with O'Toole's statement that it was five years after first being informed of the paper's serious deficiencies that he took action. He returned to the refrain of O'Toole's critics that her charges changed over time: "The serious allegations contained in the OSI report first became known to me in March this year, when the report was sent to me for comment. Five years ago, 'the problems' identified by Dr. O'Toole were of an entirely different nature."

Baltimore maintained that when they discussed the possibility of O'Toole's writing to a scientific journal about her concerns, he said the co-authors would probably respond, an appropriate action in "normal scientific disagreements."

O'Toole's contention that he believed there was no need to correct false claims because other scientists would discover and clear up the errors was "a gross parody of my stated belief that the scientific process is the best means to test the scientific validity of published claims.

"While Dr. O'Toole has now directly attacked my honesty and integrity, none of my previous remarks nor any of the remarks in this statement were intended to criticize her personally, impugn her abilities as a scientist or question her motives," Baltimore said. Nonetheless, he noted that he made his statement "in the hope that any assessment of the validity of her comments will be a measured one, based upon a consideration of all of the facts and the entire record of this controversy, including Dr. O'Toole's own previous statements on the matter."

Baltimore's renewed campaign against O'Toole was quickly joined by MIT's Herman Eisen. In the May 30 issue of *Nature,* Eisen noted that O'Toole had told him in March or April 1986 of experimental and personal difficulties she was experiencing in the lab, but he denied that she ever said she was being pressured by Imanishi-Kari to misrepresent her own results. "Indeed, Dr. O'Toole gave an entirely different version of what she reported to me in her testimony to the Dingell subcommittee on 4 April 1988," Eisen wrote. He added that she never asked for his help in finding another job, and he explained that she knew from the outset that her postdoctoral fellowship at MIT would end May 31, 1986.

These assertions were almost benign when compared to two statements in Eisen's *Nature* piece that inflamed O'Toole's supporters at Harvard. The first was Eisen's criticism of O'Toole for not charging fraud at the beginning. He claimed that her memo to him contained no hint that the paper's published results were based on nonexistent data. He was puzzled, he said, by her later charge of fraud.

Eisen said that only recently had he noticed a "hint of an explanation" when he realized that in her 1988 testimony before the Dingell subcommittee O'Toole said that her memo had been edited to eliminate language implying scientific misconduct.

"Thus it is not inappropriate to ask: Who misled whom?" Eisen wrote. "It is ironic and sad that, instead of recognizing that she bears some responsibility for creating a misleading situation, Dr. O'Toole now characterizes the initial inquiry at MIT as a 'cover-up.' Given her choice of words, I also find it remarkable that those of us who were involved in the inquiry are accused of slander and libel."

Eisen asked why O'Toole did not testify in 1988 that Imanishi-Kari had admitted that a large series of experiments had not been performed as published, if that were the case. Apparently, he did not notice that O'Toole's memo said that "the hybridomas of Table 2 were not checked for isotypes other than mu according to Dr. T. Imanishi-Kari and Dr. M. Reis." Nor did he seem to remember that he himself testified before Dingell's subcommittee in 1989 that he "was aware quite distinctly of the possibility that it [scientific misconduct] deserved fair consideration without her having to charge it." Indeed, his own December 20, 1986, memo on the controversy referred to "allegations of misconduct by Thereza Imanishi-Kari" and started off saying: "The allegations of misrepresentation were brought by Dr. Margot O'Toole. . . ." Now Eisen was standing his own testimony on its head.

The second statement that enraged O'Toole's supporters at Harvard was his explanation of why he disagreed with O'Toole on the science, particularly "her assertion that some of the results could be explained by heterodimer formation, which I take to mean mu-gamma chimaeric molecules, is highly implausible. . . . It is unfortunate that she mistakes such disagreements for slander and libel." Imanishi-Kari and her supporters had also put forth this argument, that O'Toole had interpreted the experiments' results differently and proposed a mu-gamma hookup. While the isotypes of different antibodies sometimes link up, scientists generally believe that mu and gamma isotypes cannot do so, so any implication that O'Toole suggested such a scenario would hurt her reputation as a scientist. (Ptashne would later argue that point with Eisen at an unusual meeting at Harvard requested by the MIT scientist.)

Among his other disputes with O'Toole, Eisen also supported one of Baltimore's contentions: that to maintain that the seventeen pages provided obvious evidence of problems in the *Cell* paper "grossly oversimplifies the complexity of the material." Eisen noted that Walter Stewart had found the pages confusing and that when he asked Stewart how long it had taken him to make sense of the documents, the NIH scientist replied, "Weeks." Eisen, however, seemed to take the seventeen pages more seriously than did Baltimore, who dismissed them as irrelevant.

Eisen took offense at O'Toole's reference to the Imanishi-Kari data relating to Table 3 that were in his grant application, "as though I were somehow responsible for some unsavoury data." He said that in such applications, each investigator is responsible for his or her own data, and "the data in question would have appeared only in Dr. Imanishi-Kari's portion. . . ."

Eisen praised O'Toole's tenacity, but he added that "her attacks on those who disagree with her are unwarranted."

■

Nor was that the end of the matter.

On May 30, *Nature* published what it referred to as "a much abridged version" of Imanishi-Kari's formal response to the OSI draft; the *Nature* version ran about a page and a half of its pages, while Imanishi-Kari's response ran forty-five pages, not counting additional documents from her lawyer.

Imanishi-Kari began by noting that she was forced to respond without having had access to the full Secret Service reports and data or to the original lab documents. Nor had she been given the opportunity to cross-examine O'Toole or other critics. "Most outrageously, until a few weeks ago, when the draft report was issued, my defense was conducted without benefit of knowledge of any of the specific accusations against me," she said. "I cannot disprove any of the Secret Service allegations without access to the original laboratory notebooks and tapes. . . .

"This denial [of access] should be seen in the light of the

statement of Dr. Suzanne W. Hadley . . . as quoted in *The New York Times,* May 3, 1991: 'If we showed a scientist all the evidence and let him cross-examine all the witnesses before we wrote the draft report, we would run the risk of giving him information that he could then use to cover his tracks.' Dr. Hadley fails to state the obvious, that under her unique rules the accused is denied the means to establish innocence. A more cynical person might also conclude that under these rules the OSI is able to concoct a case without fear of meaningful challenge."

(Hadley later explained that Imanishi-Kari was supposed to get her original laboratory notebooks after the draft report was issued and before it was revised and sent to the Office of Scientific Integrity Review for further action, but the documents, as well as the Secret Service data, had been impounded by the U.S. attorney's office.)

Imanishi-Kari attacked the forensic and statistical analysis, calling it "the weakest of all possible forms of evidence, as it is hocus-pocus, unproven, hand-waving, with poor scientific methods of analysis and extrapolated conclusions from flimsy, experimental results." She also argued that the so-called yellow and green tapes came in many shades.

She wondered why OSI would dismiss all the statements by her, Reis, Weaver, Baltimore, Wortis, Huber, Woodland, Eisen, and others. "It is as if the author(s) of the draft report believes all of these scientists are part of some giant conspiracy and the only person whose testimony has credence is Margot O'Toole," Imanishi-Kari said.

Imanishi-Kari disputed the suggestion by O'Toole and the drafter of the OSI report that she had written up "new data after the allegations surfaced. In fact, at a lawyer's suggestion, all I did was to take pages already written and organize them in binders in a coherent fashion. I did not write new data," she insisted.

Regarding the allegations about the June subcloning data, Imanishi-Kari contended that O'Toole had misunderstood her. True, she had told O'Toole that no isotyping was performed on those supernatants, but that was not the same thing as saying that there had been no subcloning experiments done on Table 2 hybridomas, as O'Toole claimed. Imanishi-Kari noted that O'Toole's 1986 memo to Eisen referred to the isotyping issue and made no mention of the absence of Table 2 cloning. (O'Toole later explained that there was no reason to bring up the issue of cloning at that time; she did not know until she saw the first NIH report that anyone had claimed that the June subcloning hadn't been done.)

Imanishi-Kari also denied that she had made up subcloning data. She couldn't have, she said, because she hadn't known in advance that the question would come up.

In general she was totally dismissive of all the charges.

"There are no experimental results in the *Cell* paper which can be directly contradicted by published experimental results," she insisted in her full response to the OSI report. "Interpretations of the results, however, are quite controversial. Controversy is, however, the most important part of the process of generating novel scientific findings. Controversy generated by experimental results can only be resolved by further experiments, not by the kind of debates generated during these five years of investigation!"

The OSI report, Imanishi-Kari said, "suffers from its overly ambitious effort to transform possibility into proof. No reputable scientific journal would accept a paper with conclusions based on such flimsy evidence."

■

In the June 13 *Nature,* Wortis, Huber, and Woodland responded to O'Toole's broadside with their own account of their meetings with Imanishi-Kari and O'Toole. In quoting a statement from Imanishi-Kari to O'Toole, they also raised the issue of O'Toole suggesting a mu-gamma heterodimer formation. They insisted that they reviewed cloning data for Table 2 because of a question about whether the wells had been sampled too early.

"We did not tell Dr. O'Toole that the paper 'would be retracted,' nor did we agree that it should," they wrote. "We did not tell Dr. O'Toole that 'the protection of careers must take precedence over scientific accuracy.' " They added that they never slandered or libeled O'Toole.

Not surprisingly, O'Toole could not let the issue go. In the June 27 issue of *Nature,* she challenged Eisen, insisting once again that some experiments had not been done, and reminding him, lest he had forgotten, that his December 1986 memo referred to "allegations of misconduct."

O'Toole said that if the word "fraud" had not been used in her memo, it was only because an MIT administrator had told her to remove it. And it was Eisen's job, she said, not hers, to determine if there was any wrongdoing.

Eisen's explanation that he did not realize that there might not have been data for the published experiments other than those in the seventeen pages was no excuse for the failings of his investigation, according to O'Toole. "By his own admission, he did not see or even ask to see any other data for the challenged findings," O'Toole said. "Stewart had to start by learning the meaning of idiotype and other basic concepts. It still took him only weeks to conclude that the published table was wrong."

She also pointed out that her knowledge of experiments not done, or done but misrepresented, came in large part from the meetings with the Wortis group and Eisen, before any forensic or statistical analyses were done. Said O'Toole: "Now that the draft OSI report has supported my position, it should be growing clear that the description of how I obtained this detailed and accurate information must also be correct. . . . As each piece of evidence is uncovered, someone usually has had to change his or her account to make it fit the evidence, but I have never had to do so. That my account has proved consistent with all the evidence that has come to light is no accident. I have been telling the truth all along."

The June 27 issue of *Nature* also provided vindication for Stewart and Feder. Almost four years after it was finished, *Nature* published, as a matter of record, the NIH scientists' manuscript on the seventeen pages.

Imanishi-Kari also weighed in, in the June 27 issue, with a recounting of her position and a denial that she had admitted that crucial experiments were performed. She argued that while the paper mistakenly reported that many Table 2 clones were gamma2-b, they were indeed non-IgM. In other words, she claimed, they were nonetheless of a nontransgene isotype, although that isotype was not fully characterized.

Imanishi-Kari also continued to criticize the Secret Service findings, noting that the agents said they "could not find any 'green' counter tapes in other people's notebooks after January 1984. . . . I can conclude only that the Secret Service did not look hard enough."

Imanishi-Kari's position continued to be legalistic; she claimed she had been denied due process in the investigation. She drew support from a signed statement by one hundred forty scientists, who said they were deeply disturbed that she did not have the chance to adequately defend herself.

Among the signers were former Baltimore postdoc Fred Alt, who had been appointed to the first NIH review panel and then removed after complaints about the appearance of a conflict of interest; Bernard Davis; Leonard and Leonore Herzenberg; David Parker; Joan Press; Erik Selsing; and Klaus Rajewsky, Imanishi-Kari's mentor in Germany.

The debate continued unabated in the August 15 issue of *Nature*. Once again O'Toole was not happy. She questioned why, if the Wortis group had been shown data supporting the claims in Table 2, they had not informed the NIH of this in 1987. O'Toole also wondered about how Imanishi-Kari quickly produced evidence answering O'Toole's questions in 1986, but two years later did not produce or mention that evidence to the NIH panel until the day after her interview.

Some of the most intense debate on the controversy occurred, however, not in *Nature,* but at a private June meeting with the "Harvard Mafia" requested by Eisen.

■

In the midst of the acrimonious debate going on in the pages of *Nature* magazine and in the laboratories of Cambridge, Boston, New York, and Stanford, MIT's Herman Eisen, described by other scientists as "everybody's friend" and "a bit of a schmoozer," tried to turn the heat down. On June 4, he traveled the mile separating the MIT and Harvard campuses and made his way to the department of biochemistry and molecular biology at Harvard, where the department chairman, Stephen Harrison, convened a meeting—at Eisen's request—of Walter Gilbert, Mark Ptashne, John Edsall, and Paul Doty. The four Harvard professors were considered by supporters of David Baltimore and Thereza Imanishi-Kari as the leaders of a pro-O'Toole clique that had come to be known as the "Harvard Mafia."

Although conceived by Eisen as a conciliatory gathering, the meeting lacked the air of ordinary scientific collegiality: No sooner had all the participants taken their seats than two of the Harvard scientists pulled out tape recorders.

While Imanishi-Kari was the focus of the OSI investigation and allegations of fraud, the great tension that developed between MIT and Harvard scientists was over David Baltimore's role in the controversy. The Harvard scientists were speaking publicly and quite critically of Baltimore's handling of Margot O'Toole's challenge. The MIT scientists were bitter and angry, believing that the Harvard group was being unfair to Baltimore. Underlying the surface issues was a suspicion on the MIT side that the Harvard scientists were joining the attack on the Nobel laureate out of professional jealousy.

Gilbert would not be the only one to compare the divisiveness in the scientific community to that caused by the Dreyfus case. (In the 1890s, French families split over their belief in the innocence or guilt of Alfred Dreyfus, a Jewish captain in the French army, who was wrongly, and in part because of officials' anti-Semitism, convicted of treason and sent to Devil's Island.) Gilbert lost longtime friends from MIT.

Personal and professional relationships among the Harvard and MIT scientists became so strained that surprised gossip arose when MIT's Phil Sharp, who had urged scientists to write Congress on Baltimore's behalf, not only attended the 1992 celebration of Gilbert's sixtieth birthday, but also stood up and spoke of their work together in establishing Biogen, Inc. Sharp later explained that while he did not like what Gilbert had been saying about Baltimore, he recognized that scientific misconduct had been an issue of long standing for him. He was willing to forgive Gilbert because Wally had been consistently interested in the issue of scientific fraud throughout his career. As for other critics, Sharp was more skeptical of what he sarcastically referred to as their "lofty sentiments."

Imanishi-Kari and her supporters suggested that the attacks arose, at least in part, from the intense debate within the scientific community during the 1980s over government funding of the international Human Genome Project. This multibillion-dollar research project was designed to map all of the genes—some 55,000—in a human body and determine their function. Supporters of the project—scientists like Walter Gilbert, an early developer of sequencing techniques, who said that "the human genome is like pursuing the Holy Grail"—believed that the knowledge would lead to the ability to control the function of the genes. Opponents—and Baltimore was at first one of the project's leading critics—were overwhelmed by the magnitude of such a project and feared that it would both divert federal support from other worthy research and encourage political control of science.

The argument, according to Imanishi-Kari's supporters, was that the acrimony over the Human Genome Project had spilled over into the debate on the *Cell* paper; the project's promoters had taken their fight against Baltimore and other opponents to a wholly different arena.

It was a theory that was impossible to prove. Its biggest chink was that Baltimore had eventually been won over to become a supporter of genome research. Also, there were Harvard critics of the genome project who had rallied to O'Toole's side. Rather than being tied to a specific policy dispute, the division was probably more natural. MIT and Harvard professors had been highly competitive rivals for a long time—after all, MIT had been founded, two hundred years after Harvard, by a geologist angry because Harvard wouldn't give him an appointment.

While Harvard, historically an institution for the elite, sported a handsome campus of brick walkways and elms, MIT has been generously described as "a hodgepodge of an industrial landscape bisected in several directions by railroad tracks." Nonetheless, much to the dismay of the Harvard community, it quickly became known as one of the best engineering schools in the country. More than twenty Du Ponts were educated there, as was a son of Thomas Edison; Alexander Graham Bell conducted experiments at the MIT labs.

As the years went by and MIT became wealthier, its curriculum expanded, often competing directly with Harvard's. In no area was the overlap more obvious than in basic and biomedical research.

The cultures of the two institutions remained quite different. Harvard professors were known for keeping their distance from each other and maintaining their own counsel and research agendas; MIT professors were known for their collegiality and for their devotion to instruction and the laboratory.

It came as no surprise to the MIT scientists that among Baltimore's leading and most vocal critics would be Harvard's Gilbert and Ptashne. Walter Stewart, the NIH whistle-blower, was a former Harvard student; he put it a different way: "I knew it [the paper's problems] would come out. When three Harvard scientists know something, how are you going to cover it up?"

Edsall noted that the June 4 session with Eisen was the first meeting on the Baltimore case of the alleged members of the "Harvard Mafia."

Eisen told the Harvard scientists that there were two issues on his mind: how Margot O'Toole was treated and the MIT inquiry, which he conducted. He wanted to impress upon them that while he might have made some mistakes in the inquiry, there was no deliberate cover-up of fraud.

He began by apologizing for being emotional; he was upset by statements in the press about O'Toole being fired from her job at MIT because of the questions she raised about the *Cell* paper.

The claim that she was fired was simply not true, Eisen assured the scientists. Her appointment, he said, had been for one year and one year only. When the blow-up between O'Toole and Imanishi-Kari occurred in the spring before the Wortis inquiry, it was only a couple of weeks before O'Toole's fellowship was to end anyway, he said.

Edsall interrupted, according to a transcript of the meeting, "So she had only two weeks to go anyway?"

"And so I don't know whether Margot went back to the lab or not. . . ." Eisen continued. "But her appointment ran its course those last two weeks. . . . The matter of losing her career, which also is a matter of concern to everybody, including myself, I don't really know what happened there. She says she came to me for help. I don't have any recollection of that at all. She never asked me for a letter. I think I would have written a letter for her that would have been limited to the fact that I knew her primarily from the memo she wrote, which was a very good one and I could have said very good things about her based on that. But I couldn't do more than that because I had no idea how she functioned in a laboratory."

Eisen said he believed that O'Toole's exile from science was in a way self-imposed because she felt people were hostile to her. "I think it's an exaggerated assumption and . . . I think it's sad," Eisen said. "But I don't think it's right to say that she was fired and that we kept her out of science."

Ptashne wasn't convinced. O'Toole had told him that she had had a position at Tufts lined up, but that pressure was brought against her hiring because of Eisen's memorandum saying she had alleged misconduct. "She says there was pressure brought. I don't know," Ptashne said.

Eisen said he didn't either.

But whether O'Toole lost her career was not the major issue to Ptashne. He felt that Eisen had insulted her intellect and had left the impression in the scientific community that she was "dumb."

"In the thing you just wrote in *Nature,* you said that her—and it's exactly what David said to me three weeks ago—you said that her charge was that there was mu-gamma heterodimers. Now this makes her seem very stupid," Ptashne said. He was referring to scientists' belief that while isotypes of different molecules can link up, mu and gamma isotypes cannot. Mu isotypes from different molecules can link up, for example.

"That's what she implied," Eisen responded.

"No, no, but Herman, I've gone back [and] read carefully her memo. You've studied this now for six years, so you know what I'm saying is true," Ptashne said. "She said the experiments of Figure 1 and Figure [Table] 2 could be explained by heterodimers. Those are mu-mu."

But Eisen said that the IgM molecule that she had been referring to was not a heterodimer.

Still, Ptashne continued: "Herman, you said mu-gamma. That's different from saying mu-mu. . . . You imply that she's saying that the double producers are mu-gamma heterodimers. And it makes her out to be dumb when she's not."

"No, I don't," Eisen insisted. "First of all, I've always said, and I said in that piece in *Nature,* that I found her memo to be thoughtful and cogent. That she says heterodimers, she says not once but twice. I'm accustomed to giving exams to students and grading them and somebody tells me that a chimeric molecule with two mixed heavy chains, mu chains, is a heterodimer, then they are certainly being, uh, very fuzzy."

In the end, Eisen told the group, he didn't know whether Imanishi-Kari had fabricated data or not.

"I know in the newspapers she's been tried and convicted," he said. "And in public opinion that's been true. But I don't know that there was fraud committed . . . and I don't know yet and I won't know it until I have really seen the evidence side by side with the notebooks."

He said that he had looked at the OSI report, but that the accounts of the tapes and charges were difficult to understand and he had not yet been able to evaluate them. Eisen had begun to suspect the possibility of fabrication only when he heard the first Secret Service testimony at the 1989 Dingell subcommittee hearing. But even so, he said, "I still don't know that there is. And I won't make up my mind until I've seen all the data."

Eisen admitted that he was looking at the science of the *Cell* paper and not for fabricated data. "Because I believe that if somebody thinks fraud, they don't do quite what Margot did. I suspect Margot may have been thinking fraud but she must have been very unsure" and instead, wrote a memo pointing out errors in interpretation.

"That's why she went through this elaborately constructed memo, beautifully constructed memo, in fact," Eisen said. "But you know it's as if somebody told me . . . they thought cold fusion was based on fraud. And then they go and construct a whole new theory to explain the observations that were reported."

Ptashne asked him if he believed all the data in the paper, but just had a different interpretation.

"I believe[d] all the data in the paper," Eisen replied. "I don't know now, there's a cloud over a lot of it by virtue of the OSI report, and I don't know yet how to deal with that cloud."

Gilbert acknowledged there probably was some solid molecular biology behind the *Cell* paper, and he said that he believed, along with Ptashne, that the paper probably was not "fraudulent in its inception." But he did believe it involved misconduct in the sense of tremendous carelessness.

"My actual belief about the paper," said Gilbert "is that the immunology was so badly done that [it] has to be just wrong. And that the immunology was used to convince the molecular biology group the cells are behaving in a certain way. And the molecular biology was actually done badly and used to convince Thereza that things were working. And you have a typical case in which both sides get caught in the trap. . . . So they're happy to say, 'Well, if I see a trace of mu RNA, it doesn't matter because I'm told there's no mu protein.' "

Despite the tension of the meeting and the MIT scientist's feeling at times that Ptashne was misrepresenting his statements, the group was also able to laugh. At one point, when Eisen said he had not read MIT official John Deutch's testimony from the 1989 congressional hearing, Ptashne said: "Really? You were sitting next to him when he gave it."

"That doesn't mean I know what he said," Eisen replied to laughter and responses of "Fair enough. Fair enough."

When the Harvard group pressed Eisen about his December 30, 1986, memo, the one that referred to allegations of misconduct, he said, "I was not using 'misconduct' in the sense of fraud, of fabrication of data. I wish that word 'misconduct' had never been invented. . . . Frankly, I always thought the real misconduct in that case was O'Toole being harassed. But that is something that never really surfaced and I didn't want to talk about it."

The comment brought looks of surprise to the faces of the Harvard men. It didn't track. Eisen had just written in *Nature* that he was unaware of pressure being applied on O'Toole to misrepresent her own results, and he had specifically denied O'Toole's contention that he had failed to help her. He was asked to explain the inconsistency.

Eisen attempted to explain that he was referring to O'Toole's statements to him that Imanishi-Kari had been upset by the fact that O'Toole's experiments weren't working and had ordered her to stop the experiments and take care of the mice. "That's quite different than saying that she was being pressured by Imanishi-Kari to falsify results. . . . That's what she put . . . in her written statement to the Dingell subcommittee. . . ." Eisen said. "I would never tolerate somebody coming to me and saying, 'My supervisor is telling me to fabricate data.' "

Ptashne pointed out that the same written statement to Dingell's panel in 1988 also included O'Toole's contention that she had been pressured to misrepresent her results. In fact, O'Toole's written statement in 1988 said she had informed both Eisen and Wortis of the difficulties that she was facing with Imanishi-Kari and that Eisen and Wortis had refused to get involved. But Eisen reiterated that O'Toole had not made that actual charge to him in 1986.

Ptashne was skeptical.

Eisen retorted: "You're assuming that Margot O'Toole's memory is infallible. I assume . . . Margot O'Toole, like anybody else, will remember things in different ways. . . . I don't trust her memory and I don't trust my own."

Ptashne replied, however, "The problem we have with O'Toole's memory is that [on] every scientific point that can be checked, she turned out to be right. When she saw that subcloning data, she said it was fraudulent. That's the first time she used 'fraudulent.' She said if you look at the notebooks you'll find that it didn't exist originally. She recited chapter and verse."

But Eisen wasn't impressed with notebook examinations. He pointed out that he had been involved in investigating earlier allegations of fraud at Harvard Medical School and in that case the questioned notebooks appeared perfect; the fraud came out because the accused confessed. "So, I don't have the kind of respect for notebook inspection" that Walter Stewart does, he said.

Gilbert agreed that catching fabricated data was difficult "because the most fraudulent cases are the most elegant notebooks." But by checking the notebooks, he explained, "you catch a different kind of problem. You catch, actually, a more common one, which is self-deception."

Doty, who would have to leave soon for a doctor's appointment, interrupted to raise the issue with which he was most concerned in the controversy: the way science should be conducted, fraud questions aside. He contended that the co-authors had strayed in this case from following the "Feynman principle."

■

Considered one of the few true geniuses of our time, the late Richard Feynman, who won the Nobel Prize in physics, articulated an underlying principle for the conduct of scientific research: A scientist should explain what is wrong with his experiment, not just what is right and will prove his results.

"Details that could throw doubt on your interpretation must be given, if you know them. You must do the best you can—if you know anything at all wrong, or possibly wrong—to explain it," Feynman said in a reminiscence adapted from his 1974 commencement address at the California Institute of Technology. Maintaining this "utter scientific integrity," he said, would help scientists avoid the trap of fooling themselves.

Scientists speak reverently about the Feynman principle, and Doty clearly felt that Baltimore and his collaborators had not conformed to it, although, he acknowledged, "We know it was not their practice in previous work." The opposite of Feynman's code of scientific conduct, Doty said, "is to publish whatever you can get away with. I think that the totality of this case does indicate that the authors have migrated pretty far toward this other end. . . .

"And it's in that sense that I thought David's mea culpa . . . was only about one-third of the answer," Doty said. "It was withholding the essentials of the argument that disservices science all the way around."

While Eisen said he agreed with the Feynman principle, he continued to maintain that the whistle-blower bears some responsibility to make a clear allegation of fraud.

This infuriated the Harvard scientists. Gilbert insisted that it was the responsibility of senior scientists to think about the possibility of fraud; a junior scientist such as O'Toole should not have to use the word "fraud."

Doty agreed. "If we follow your injunction, then no whistle-blower would ever blow the whistle until he or she was certain it was fraud, provable fraud. And that, that negates the whole possibility of useful whistle-blowers."

Ptashne believed that Doty had hit on the big issue: the integrity of science itself.

"I hope you realize that our problem is, I think Paul really put it very eloquently, is the larger one," Ptashne said. "And to the extent that there's pressure on us to keep quiet and not talk about it, to make it go away, I mean, we're in trouble. . . . We grew up reading Feynman, and Feynman . . . says we have our individual responsibility."

Ptashne was particularly upset with Baltimore's stance that it was up to other scientists to prove or disprove the *Cell* experiments. "I called David and I said, 'David, how can you say this?' He said, 'I was misquoted.' I then find he said that in the hearings in '89. . . . I don't think ultimately any of these institutional responses mean diddly-squat. . . . It has to do with how individual scientists act."

"But they're not getting, from David, Feynman's message," Gilbert added. "They're getting from David . . . the message that this is an attack on science."

Eisen said he agreed with Feynman, but "I don't think we can be expected to act like lawyers."

"No, you've missed the point," Ptashne lectured him. "It's acting not like a lawyer. It's acting like a scientist who will get to the bottom of his own problems. That's the message we have to give to the world. Otherwise . . . they have to have a policeman in every laboratory; they'll destroy us. And far from this being an attack on science, the only chance that we save ourselves is by making this ethic clear."

If that was the case, what should scientists do? asked Eisen. Was Baltimore's handling of this situation so devastating, such a departure from the Feynman principle, that unless science as a whole publicly proclaimed it, the policymakers in Washington would abandon science? Was it so devastating?

Gilbert nodded; he thought Baltimore's actions were, indeed, a departure from Feynman.

Eisen was clearly becoming more troubled. Ptashne, who had gone head to head with Eisen, tried to cheer him up.

"I think everybody at the table is sympathetic to the spot you were in," Ptashne said. "So it's not a personal thing. . . . But we can't be sympathetic to your saying . . . that somehow everything was okay and the problem was Margot's. That puts the scientific community in the anti-Feynman mode."

"I think that you are stating my position in a way that is extreme and I don't agree with it," Eisen responded. "Wally says the validity of the science is irrelevant. I think to some extent he is right, now, in retrospect. It was not the way it appeared to me then. There's a difference between the way things looked five years ago and the way they look now. And it's easy in hindsight to point out all the shortcomings of what we did."

Eisen couldn't bring himself, however, to condemn Imanishi-Kari. "I'm waiting to see the full significance of Imanishi-Kari's response," he said, "I want to see her given a chance to respond to the Secret Service data. And I think that till I've seen that I . . . want to leave the chink in my mind open to the possibility that there was no fabrication of data. . . .

"I mean basically we're asking to condemn her, to declare a verdict of murder. This is the equivalent of murder," Eisen said. "And I will not, I will not vote for a murder conviction unless I really have evidence that I find absolutely incontrovertible."

But Ptashne couldn't help wondering what the effect on science would be if Imanishi-Kari were to be indicted and convicted in a public court of law. "It's not a question of our condemning her," he said. "It's a question of what's happened."

"We can't look much worse than we look already," Eisen replied.

The meeting that Eisen had initiated with hopes of ending the feud concluded with all sides seeming more glum than they had begun. Little had been settled and nothing anyone tried could make the matter go away. Even this session couldn't end without a last-minute return to the debate over whether O'Toole had impugned Eisen's honesty and whether he had gone overboard in his *Nature* response. Everything in the scientific community was subject to debate and redebate. The scientific principle of experimentation and reexperimentation appeared to be applying itself to their political discourse as well. Nothing could ever be settled. Each exchange gave way to a new exchange, then a reexchange, and so on.

Eisen and Ptashne would later finish their debate over heterodimers in an exchange of polite letters in *Nature.* And Paul Doty would find the need to expand his belief that David Baltimore had not acted properly as a scientist. The session had convinced him that this was an important point that needed to be aired.

It would be Baltimore's response to Doty that would seal his fate at Rockefeller.

SEVENTEEN

Rough Justice

■

Herman Eisen's meeting with the Harvard scientists clearly did not have the effect of quieting the controversy, although he and Mark Ptashne did have a civilized exchange of letters in the July 11 issue of *Nature*. They discussed O'Toole's contention that the positive idiotype results were the result of heterodimer formation, molecules including mu chains of both the transgene and the mouse's own (endogenous) gene.

In his letter, Ptashne first described how two other related papers did not replicate the central claim of the *Cell* paper. He wrote, for example, that a 1989 paper by Satyajit Rath et al. in the *Journal of Immunology* showed that mice engineered with a different idiotype than the one used in the *Cell* paper expressed antibodies with the transgene idiotype but "all detectable AD8 reactivity [transgene] was associated with molecules expressing the mu^a allotype and none was detected in association with molecules lacking mu^a." Ptashne then said that, contrary to Eisen's earlier *Nature* piece, O'Toole was referring to the possibility of mu-mu chimeras, not mu-gamma.

"O'Toole's explanation, according to Rath et al., is eminently reasonable and readily testable," Ptashne wrote. "I thank Herman Eisen for graciously reviewing this matter with me."

In his response, Eisen noted that he had not realized that O'Toole had meant a "mixed IgM molecule . . . in which some of the mu chains were encoded by the transgene and others by endogenous mu genes. In contrast to heterodimers, such mixed mu-mu heteropolymers are certainly plausible and O'Toole's suggestion would thus appear to have been reasonable."

But Eisen added that even if there had been no misunderstanding, he would have believed that more research was necessary to evaluate O'Toole's theory. He also discussed Rath et al. and the second paper, one by Jeannine Durdik et al. in the 1989 *Proceedings of the National Academy of Sciences USA* that Baltimore said supported the *Cell* paper but one of whose authors, Erik Selsing, said did not provide evidence of idiotypic mimicry.

"Besides being elegant studies in their own right, Rath et al. and Durdik et al. are welcome demonstrations that the scientific process itself is the most effective way of resolving scientific disputes," Eisen wrote, adding, "I look forward to additional research that bears on disputed scientific issues in the *Cell* paper."

So Eisen made amends on the heterodimer dispute, but he still maintained his outlook on the controversy as being a dispute over scientific issues, not over whether experiments were done or not done as published.

However, while Ptashne and Eisen were cooling off, other scientists joined in the debate in the very same issue of *Nature*. John Cairns of the cancer biology department at Harvard's School of Public Health sent in a copy of a letter he had written to an unidentified officer of the National Academy of Sciences, in which he described the *Cell* affair as "a kind of scientific Watergate." Cairns originally had supported Baltimore, but after reviewing the case had come to believe in O'Toole.

"Nothing now is likely to stop the affair from progressing to its final disastrous conclusion, and I do not see how the authors of the paper can escape public censure at the very least," Cairns wrote. "About the only question remaining is whether anyone will actually go to jail."

Cairns laid some of the blame on the scientific community and "the many scientists who did not look at the evidence and, instead, construed the whole business as a congressional maneuver to attack the scientific establishment. (I remember that originally I, too, felt that the row was probably a political stunt.)"

Nicholas Yannoutsos, who worked with Imanishi-Kari from October 1985 to May 1988, provided a distinctly different perspective. His work, he said, was based on the *Cell* findings, which were a "part of continuing research conducted with genuine and critical interest. I must stress the 'critical interest' because Imanishi-

Kari, herself, and other people in her laboratory, including myself, have painstakingly repeated time and again the work reported in the *Cell* paper. . . . All this was done in an atmosphere of openness and intellectual integrity."

Yannoutsos defended Imanishi-Kari's work and said that she demanded strict attention to the technical aspects of the experiments. "Her strictness might occasionally frustrate the false pride of an individual worker, but it was also a lesson in the essential modesty, dedication, and correctness with which scientific work must be conducted and of which Dr. Imanishi-Kari was herself the best example."

Paul Doty also was not satisfied to leave his actions to a private discussion with Herman Eisen and his colleagues at Harvard. He felt he had a duty as a scientist to take a public stand on David Baltimore's role in the *Cell* controversy. This feeling of obligation came as no surprise to people who knew him.

Doty, seventy-one at the time, was director emeritus of the Center for Science and International Affairs as well as the Mallinckrodt professor of biochemistry at Harvard. A native of Charleston, West Virginia, Doty earned his doctorate in physical chemistry from Columbia University in 1944 and worked at the Polytechnic Institute of New York and the University of Notre Dame before being named a Rockefeller fellow at Cambridge University. He moved to Harvard in 1948. Some of the best-known work of his laboratory involved the molecular melting of DNA and its renaturation, which underlies much of recombinant DNA technology. Doty was one of the founding editors of the *Journal of Molecular Biology* and the *Journal of Polymer Science*.

Originally a member of Harvard's chemistry department, Doty had worked on the top-secret Manhattan Project during World War II. Later he helped establish the department of biochemistry and molecular biology at Harvard and served as its first chairman. A program he began at Harvard in 1973 eventually became the Center for Science and International Affairs; he was director from 1973 to 1981. When Doty retired from biochemistry in 1988, he became a professor of public policy at the John F. Kennedy School of Government at Harvard.

In the July 18 issue of *Nature,* more than a month after meeting with Eisen, Doty published a commentary about Baltimore's actions "in the hope that it might stimulate others to examine the matter." More important than the actual validity of the science,

Doty said, was whether the account by Baltimore and his colleagues "represented a departure from the normal standards of research, which puts the uncompromising search for truth first."

Doty was not satisfied with Baltimore's apology, which he did not believe went far enough. He noted that it came "five years after he [Baltimore] was informed of serious shortcomings in the work and two months after severe condemnation in the draft report of OSI." Doty acknowledged that the final NIH report might differ from the leaked OSI draft, but he believed that "the case for egregious departure from the usual standards of carrying out and reporting research stand independent."

First, Doty wrote, Imanishi-Kari's recording of data "was so sloppy as to insult the scientific method." Then, Baltimore failed to examine critically the quality of the data before publication and then after serious criticisms were raised; he did not further test the conclusions; and he "organized an attack on his critics." One example of the co-authors' departure from what should be normal scientific operating procedure, he said, was the publication of Weaver's Figure 4, defended by Baltimore at the Dingell hearing, without showing the weak band of the transgene.

Doty blasted Baltimore's statements that it remained for other scientists to prove or disprove the *Cell* experiments. "To forgo this obligation—to leave to others the responsibility of establishing the validity of what you have published—is not only a fundamental retreat from responsibility but, if it became accepted practice, would erode the way science works," he wrote. "For the cutting edge of science moves forward by building rapidly on what is published on the tentative assumption that is correct, not by waiting for others to test each paper's validity."

Using improved techniques, more definitive data, or clearer thinking, other scientists down the road might discover honest errors in published work. That was a far different "scientific method" than the one Baltimore seemed to be backing: Publish the stuff and require others to prove or disprove it.

Doty also attacked what he viewed as Baltimore's continuing effort to conceal the whole truth, particularly his statement that O'Toole's response to his apology failed to "mention that the seventeen pages have been proven to be irrelevant." Doty noted that the OSI draft describes the way in which the data reported in the seventeen pages was presented in the *Cell* paper as "not defensible." Therefore, the seventeen pages are hardly irrelevant and it is misleading to suggest that O'Toole should have said they were.

"This pattern of behavior stands in deep contrast to the traditional view that authors of scientific papers have a special obligation to be responsive to criticism and to test their work from every possible angle—to pursue the truth relentlessly," Doty wrote.

Doty said Baltimore's actions in this case were unlike him, but "to let this lapse pass in silence would be to condone it and to fail to recognize the very different standards of his other work as well as the harm that has come from the posture he has taken."

In his commentary, Doty also criticized Tufts and MIT for not appointing properly selected committees that would have adequately and promptly reviewed O'Toole's concerns, and the first NIH panel for substituting the supplementary June subcloning data for the questioned data. He did, however, credit Eisen for correcting the misunderstanding about O'Toole's position on heterodimers.

Scientists cannot leave it to Congress or the courts to get a handle on dealing with scientific misconduct. The very nature of the problem is such that they must do it themselves, he said.

Unless the scientific community dedicates itself, as O'Toole did, to what the OSI draft called "the belief that truth in science matters," Doty said, "its silence can be interpreted as condoning the standards of research and reporting embodied in the . . . [*Cell*] paper."

He warned the scientific community that it was already facing threats from the "new stresses and temptations" of cutthroat competition for grants, corruption problems with peer review, and the growing cases of deception in published papers.

"All these contribute to the pressure to compromise and erode the high principles of the past," Doty wrote. "As a result, the scientific community may already be experiencing a gradual departure from the traditional scientific standards; this could be abetted by condoning the behavior seen in this present case.

"In this way we risk sliding toward the standards of some other professions where the validity of action is decided by whether one can get away with it. For science to drift toward such a course would be fatal—not only to itself and the inspiration which carries it forward, but to the public trust which is its provider."

■

Doty's commentary angered Baltimore.

The Nobel laureate's response, in the September 5 issue of *Nature*, was less than a column and a half long, but it did what many scientists felt was irreparable damage to his standing at Rockefeller. In a "Dear Paul" letter, Baltimore returned Doty's harsh assessment with a stern lecture that refused to recognize any failings with respect to the *Cell* paper. This letter seemed not to have been written by a contrite scientist who only months earlier had retracted the paper in question, and in apologizing to Margot O'Toole had expressed his "tremendous respect" for her.

"You have used a leaked, confidential draft of a government report as a basis for definitive judgments about the acts of others," Baltimore wrote combatively, adding that without all access to "the evidence," "your judgments therefore could not have derived from the careful assessments that you yourself say that scientific evidence requires. . . . Your judgments do not depend on complete evidence; your verdicts are, rather, based mainly on the unsubstantiated, and often refuted, allegations of one participant in events five years old."

Baltimore argued that his science, including the work of the *Cell* paper, was "done with rigor and criticality. But, because the issue is one of judging historical events, the only way to judge that statement is by the traditional test of science: Have the data proved reliable?"

Ignoring his own retraction and the reasons that led to it, Baltimore contended that "the data have proved more durable than the data in most papers. No experiments of which I am aware have appeared in the literature that contradict the data of the . . . [*Cell*] paper. In fact, there is much published evidence and more coming that support the paper's results in remarkable detail."

Baltimore said Doty's criticism of Figure 4's publication was "unworthy of you," contending that "no one consciously eliminated any band."

"I welcome your call for the highest standards in the prosecution of scientific investigations. I have supported just those values all of my scientific life and will continue to do so," Baltimore said, concluding: "In contrast to your apparent certainty about the events of the past, I am not now convinced that I know all the answers. But I do know that I have tried to ferret out the truth and I feel that I did reasonably well because the science has stood up to the toughest test of all, the test of history."

Baltimore's response to Doty stunned scientists at Harvard, Rockefeller, and elsewhere. Was he retracting his retraction? Was he actually saying that the findings of the *Cell* paper could be both valid and fabricated by Imanishi-Kari? The OSI was not accusing her of bad interpretations, but of making up data and not conducting experiments that she said were done.

Rockefeller scientists, particularly, feared what Baltimore might say next. They worried whether John Dingell might hold another hearing. Even worse, they wondered what the U.S. attorney would tell members of the grand jury meeting in Maryland. A secret straw poll of the senior faculty earlier in the spring had already indicated that only one-third supported Baltimore.

"He even retracted his retraction. . . . That's what made the faculty upset," one Rockefeller scientist told *Science*. "They said, 'We can't support those arguments.' No one can defend this position. He was saying the paper still stands up as well as any other in the literature. Do people believe that?"

Every time Baltimore's name was mentioned in the press, Rockefeller University was also dragged in. "Rockefeller has great name recognition," said Norton Zinder, the molecular biologist who, after the OSI draft had been leaked, urged Baltimore to be apologetic. "People just see the headlines [and say] 'You work at that place that has those crooks?' "

When scientists at Rockefeller and elsewhere heard Rockefeller biochemist Anthony Cerami's announcement on July 31 that he was leaving, along with his thirty-member research team, for the newly established Picower Institute for Medical Research, the first thought of many was that his leaving had to do with the Baltimore controversy. Press accounts echoed this view. Cerami, after all, was one of the prominent critics of Baltimore's appointment to the university's presidency; he had said that Baltimore's poor handling of the case was "a bellwether for his character." But Cerami, the former dean of graduate studies, insisted he was leaving only because of the "unique opportunity" that he had received in being offered the presidency of the new institute.

Around this time an informal poll of the faculty indicated that a majority wanted Baltimore to resign. In a series of meetings between trustees and senior faculty members, his critics expressed concern that the continuing controversy would undermine the university's ability to recruit scientists and raise funds.

Zinder, who had urged other faculty members to rally around Baltimore once he was appointed president, was one of those senior scientists who now believed that Baltimore should resign for the good of Rockefeller University.

"It had nothing to do with guilt or innocence, but the best interests of the university," Zinder said. "The people at the university felt they were in a war zone. Every week everyone was looking at the newspapers, *Nature,* and *Science.* People and institutions cannot live in notoriety. It was totally distracting."

The situation was not helped by the August 26, 1991, issue of *Time,* whose cover illustration showed a scientist under a microscope. The headline read: "Science Under Siege / Tight money, blunders, and scandal plague America's researchers." The article discussed scientists' malaise, tight government funding, and a catalogue of embarrassing incidents in science. Prominently included—along with the cold-fusion fiasco; the HIV discovery dispute, in which American researcher Robert Gallo finally acknowledged that the virus he said he had discovered indeed came from French scientists; and the faulty $1.5 billion Hubble space telescope—was the Baltimore case. Baltimore's photograph was placed under the banner headline "Frauds and Embarrassments."

Then, on October 8, Gerald M. Edelman, a Nobel Prize–winner who headed the independent Neurosciences Institute based at Rockefeller, announced that he and his group, along with colleague Bruce Cunningham, would be leaving the following summer for the Scripps Research Institute in La Jolla, California. *The New York Times* reported that both Edelman and Rockefeller officials said that the move was not directly related to the turmoil stemming from Baltimore's presidency.

On October 10, an unsigned column in *Nature* noted that even though the participants in the *Cell* controversy had more they wanted to say, the journal had decided not to publish more debate. The journal's "readers have seen the essence of the dispute laid out, and are likely to be confused by ever more detailed statements." But *Nature* itself had more to say about David Baltimore. This was a blow to Baltimore, since the magazine's editor, John Maddox, was a friend.

The journal proclaimed that the scientific community, not the NIH, needed to decide an issue raised by Paul Doty: "What are the responsibilities of the authors of a published research report?"

Baltimore's contention that it was for other scientists to demonstrate the validity of the disputed data and conclusions drawn from them might be a "a point of view, but hardly a defensible one," the journal stated quite starkly.

"The plain truth is that the authors of all published research reports have a personal responsibility for their aftercare," *Nature* said. "They, after all, are best placed to carry out the meticulous examination of the original data when questions are asked about them and, when necessary, to design and carry out replicate experiments."

Anything less would undermine the whole scientific enterprise, according to the journal: Scientists, unable to depend on published work, would have to repeat the experiments before doing their own work. "Naturally, it must often be a fiendish nuisance for authors to have to shoulder this responsibility, especially if many months have passed. But that is merely one of the penalties of authorship."

■

Seven days later, on the morning of October 17, 1991, David Rockefeller, who two years earlier had urged Baltimore to accept the presidency, gave the university a $20 million gift, capping its successful fund-raising effort and expressing his "absolute confidence" in Baltimore. The trustees were pleased with Baltimore's efforts to deal with the university's budget crisis and carry out reforms of the faculty structure.

That afternoon, however, they met with a group of professors who brought unwelcome news. The faculty group said there was little support for Baltimore; he had to be replaced. The trustees had asked Torsten Wiesel, also a Nobel laureate, to survey the tenured faculty in early October. He found that 70 percent of forty-four tenured scientists did not support Baltimore. The trustees then asked him to put together a group representing supporters and opponents, which he did; they met October 17, at what participants described to *Science* as a "very calm" meeting.

Baltimore, who was implementing plans to promote and hire younger scientists, began lobbying junior members of the faculty to support him before the trustees. Three members met with trustees on November 21. Some have suggested that the older scientists opposing Baltimore were taking advantage of the *Cell* controversy to attack him for aggressively promoting junior faculty and for not being more sensitive to the history and culture of Rockefeller University. Senior faculty members, however, claimed the reforms affecting junior faculty had already been approved by the faculty senate before Baltimore arrived.

On November 22, the senior scientists, angry about Baltimore's manipulation of the junior faculty, learned that the trustees had decided to continue backing Baltimore. Furlaud, the chairman of the board, went around talking to individual scientists, trying to shore up support.

It seemed as though Baltimore would tough out the crisis. On November 25, Stephen S. Hall, a free-lance writer, interviewed him for *Science* and found a "confident and upbeat" man, who contended that his position had been misinterpreted. Of course he understood the responsibility of a scientist to respond to a challenge to his work. He downplayed the significance of the October 17 meeting:

"[When I was appointed] I felt there was enough support to carry forward the plans, to do what seemed necessary to do. And that has turned out to be true. And I think that's more important than all those statements about how many people are on one side or the other. New people are coming to the campus. A tremendous amount of money has been raised; more will be raised. People are working together in a very effective effort. And we can talk about our differences, and perhaps that's as important as anything."

But on the very next day, James Darnell, Baltimore's longtime friend and colleague, submitted his own resignation as vice president for academic affairs. While Darnell refused to return reporters' calls about his reasons, scientists at Rockefeller whispered that he stepped down in an effort to pressure Baltimore to resign for the good of the university. The blow was terrible to Baltimore, who many years before had been a student in Darnell's lab at Rockefeller. Over the Thanksgiving weekend, Baltimore called Furlaud with his decision: He was resigning as president of Rockefeller.

■

Baltimore's December 2 letter to Rockefeller and Furlaud said that it was "with a sense of enormous regret and deep personal disappointment" that he was stepping down as president of the university.

"The reason I have decided to take this step is that the *Cell* paper controversy created a climate of unhappiness among some in the university that could not be dispelled," Baltimore wrote. "When I accepted the position of president of this institution, I did not anticipate that this matter would become such an extended personal travail for everyone involved. Trying to govern the university under these conditions has taken a personal toll on me and my family which I can no longer tolerate.

"Therefore I cannot lead the Rockefeller University as effectively as I would like and as effectively as it deserves to be led. Accordingly, out of my enormous respect for the Board of Trustees and the faculty of this great institution, I have decided that the time has come for me to step aside as president to devote my full attention to research and teaching."

His plans were to take up a major research project on AIDS, which he had dropped when he took on the leadership of Rockefeller.

The trustees accepted his resignation, along with that of James Darnell, at its regularly scheduled meeting on December 3.

The New York Times summed up the case in a December 5 editorial entitled "Rough Justice for Dr. Baltimore." The Nobel laureate "paid a harsh but necessary penalty for his efforts to whitewash a case of scientific fraud." While some might think that the penalty paid was out of proportion to his actions, "there is a rough justice at work here, holding even one of the nation's preeminent scientists accountable. Dr. Baltimore's fate carries a warning to the scientific community that failure to confront serious charges about the integrity of research will no longer be tolerated."

Consistent throughout, *The Wall Street Journal* on December 4 proclaimed: "Dingell Gets Baltimore": "As always with investigations conducted by Mr. Dingell, the exterminating angel of the regulatory authoritarianism movement, the victory has produced mostly rubble. . . . It isn't clear, though, that the scientific community will ever acquire a reliable set of facts around this case."

Baltimore loyalist Bernard Davis wrote in *The Los Angeles Times* on December 10: "Initially, Baltimore had wide support from scientists, because he was defending important principles. But he gradually lost support, as his courage against such a powerful adversary was increasingly seen as arrogance. The outcome has not only been a personal tragedy, but one for all of science and society. . . . If we wish to move on constructively from the Baltimore tragedy, we should now focus on setting proper limits for the role of government in dealing with scientific fraud."

Nature saw the resignation as not only a personal defeat for Baltimore but a defeat for research. The journal said there was plenty of blame to spread around: Some went to Tufts and MIT for conducting casual reviews of O'Toole's complaints; some to the NIH for taking so long in conducting its investigation; and some to the scientific establishment for not getting actively involved. "Even if the charges in the draft report are proved false, the research community will be blamed for taking so long to fail to get to the bottom of things, and for giving the whistle-blower in the case a rotten time."

But as *Nature* noted, the affair was not over. As Baltimore's story was being played out at Rockefeller, the controversy continued to rage at the National Institutes of Health, whose new director was locked in battle with the chief investigator of the *Cell* and Gallo cases. And the U.S. attorney's office in Baltimore, which was bringing witnesses before a grand jury during the spring, would throw in a wild card.

EIGHTEEN

Keystone Kops or Science Police?

■

As the scientific community debated the stunning revelations of the leaked draft and Baltimore's actions, a related controversy erupted over the Imanishi-Kari case and other scientific-misconduct cases at the National Institutes of Health. The NIH dispute bore many disturbing similarities to the Baltimore affair; at the heart was an unknown scientist convinced of her propriety in battling the scientific establishment.

Shortly after the OSI draft report was leaked to the press, Suzanne Hadley, deputy director of the OSI and the chief investigator in the Imanishi-Kari and Gallo cases, resigned to become a special assistant for legislation for NIH's new director, Bernadine Healy. Hadley, who at one point had been the acting director of the OSI, decided to change jobs because there was a growing awkwardness between her and Jules Hallum, the man who eventually received the OSI director's job. However, NIH officials agreed that Hadley would stay with the Imanishi-Kari and Gallo investigations through until the end, even as she was analyzing science policy issues and preparing legislative testimony for Healy.

Healy, who was confirmed by the Senate as NIH director on March 24, had been head of research at the Cleveland Clinic Foundation when she was chosen by President George Bush to run the NIH; the agency had been without a permanent director for two years since the resignation of James Wyngaarden. The first female assistant dean at the Johns Hopkins School of Medicine, Healy was a prominent cardiologist and an aggressive advocate for research.

She was president of the American Heart Association from 1988 to 1989. She served a stint as deputy director of the White House Office of Science and Technology Policy under Ronald Reagan, and as vice chairman of the President's Council of Advisers on Science and Technology under George Bush. Healy's tenure at the NIH got off to a tumultuous start; she quickly clashed with both Hadley and Dingell over the OSI's role.

While Healy was at the Cleveland Clinic Foundation, where her husband is chief executive, she headed a committee investigating charges that one of her researchers had not done the work he claimed in a $1 million–plus grant application to the NIH. After a perfunctory review, Healy's panel criticized the researcher but found no serious wrongdoing. He was guilty, the panel said, of "anticipatory writing." However, after the scientist complained about the criticism, Healy froze the federal funds and convened a second committee, on which she did not sit, to review the case. That panel found the scientist guilty of misrepresenting his research, and their finding led to a fuller investigation by the clinic. The third inquiry determined that the scientist had not intended to deceive the NIH.

The OSI, informed of the investigations as required by federal regulations, then looked into the case, and a preliminary report by Hadley found that the researcher was guilty of scientific misconduct in that he had made false statements to the NIH about his brain research. Hadley further faulted the panel's conclusion that the scientist was guilty of nothing more than carelessness; and her report criticized the inclusion of one of the researcher's collaborators on the inquiry panel.

Initially, Hadley did not have any problems working with Healy, who had recused herself from any NIH dealings with the Cleveland Clinic case. "We were congratulating each other on thinking alike," Hadley recalled. This amity would not last.

Hadley had become the OSI's institutional memory, having joined the office shortly after it was established in 1989. A petite West Virginian, who earned her master's in experimental psychology at Hollins College in Virginia and her doctorate at the University of Massachusetts, Hadley was a clinical research psychologist studying depression. She went to the National Institute of Mental Health in 1978 as a researcher and became the agency's misconduct-policy officer. Hadley was detailed to the new OSI and became its acting director in October 1989.

"I was so caught up in the scientific-integrity work, I was loath to leave it," Hadley said of the thirteen-hour work days and of covering her basement floor with documents to continue her work at home. "It felt heavy on our shoulders, but it felt like it mattered. It was the last best chance of the scientific community to step up to the mark and be responsible, short of science cops."

The beginning of the end probably started in May 1991 as Hadley was organizing an OSI investigation into how the MIT and Tufts people had conducted their inquiries on Margot O'Toole's concerns. Why did those inquiries find no problems? Was there a cover-up? The investigation, routine when OSI reaches a conclusion of scientific misconduct contrary to the sponsoring institution's findings, had been approved by NIH officials before Healy came on the scene.

In preparation for the planned interviews in Boston, OSI was asking officials at Tufts and MIT for any more documents that had not been turned over earlier but that might be relevant. Going through university files, a lawyer for Tufts found a letter of recommendation that Ursula Storb, a member of the first and second NIH scientific advisory panels, had written for Imanishi-Kari when she was applying for the job at Tufts. Storb was one of the second panel's two members who had issued a minority opinion agreeing that two sections of data had been made up but dissenting from the rest of the OSI report.

"I've got to have that letter," Hadley told the lawyer.

After Hadley received the letter, the OSI asked Storb to step down from the panel because she had not disclosed her relationship with Imanishi-Kari before taking on the job of investigating the *Cell* paper. But Storb refused to resign, later calling the matter "ridiculous." A specialist in the immunology of transgenic mice, Storb had been asked by Tufts for the recommendation before she joined the first panel and she did not remember it later. Storb had made her recommendation based on Imanishi-Kari's published work, hearing her speak at immunology meetings, and her résumé. Immunology is a relatively small field and Storb had at least a passing acquaintance with everybody in it.

As Hadley pursued the Tufts inquiry, Healy, only weeks after taking office, began to get involved in a series of controversies over the case of the AIDS researcher Robert Gallo, who was accused by French scientists of appropriating their AIDS virus for his patented test. (Royalties on the AIDS test bring in millions of dollars annually for the U.S. government and $100,000 yearly for Gallo.) Along with many other scientists, to say nothing of the supporters of Gallo, Baltimore, and Imanishi-Kari, Healy started to worry about the adequacy and fairness of the OSI's procedures. Then, shortly after one firefight over Gallo, the NIH's deputy director, William Raub, informed Healy that Hadley, with his and Hallum's agreement, had asked Storb to resign from the panel investigating Imanishi-Kari.

Healy was upset, saying that she was not convinced that a conflict of interest actually existed and that Storb's disqualification might require that the whole investigation be redone. Healy met with Hadley and other NIH officials May 23 about the Storb matter, and asked them to look into it further before taking any more action. She said that while she disagreed with them, she would accept their decision to ask Storb to resign.

The meeting was contentious, Hadley reported, with Healy referring to OSI as a group of Keystone Kops and charging that Hadley was "knuckling under" to political pressure from Dingell. Robert Lanman, who was also at the meeting, later said he understood Healy's Keystone Kops comment to be only a reference to a statement by Robert Charrow, the former deputy general counsel of HHS who was representing a scientist under investigation. Charrow's analysis of the OSI was being used by several scientific associations that were challenging the office's procedures.

Healy later told Dingell that her only other involvement with the Imanishi-Kari case was a decision to concur in an OSI decision to delay further action on the report because the evidence and data had been impounded by the U.S. attorney, who was holding grand jury hearings that spring, and Imanishi-Kari thus had no access to the impounded evidence in developing her response.

Hadley went to Boston the end of May for the interviews with Wortis, Eisen, and others about the early inquiries on the *Cell* paper. On June 4, she heard from Healy, who wanted the OSI report on the Gallo case to be rewritten because it read too much like a novel. Healy said the report was badly written and unclear. Hadley and Hallum refused, arguing that a change in format would change the substance of the report; Healy said she then dropped the matter.

■

On June 7, Lanman went to visit Hadley because Healy, who said she had heard from some aides that Hadley had become "too close" to Margot O'Toole, wanted him to get and review Hadley's telephone logs to determine if she had been getting an inordinate number of calls from O'Toole. Hadley directed an OSI staffer to collect her telephone notes, which were in the OSI files, and turn them over to Hallum.

Lanman later testified before Dingell's subcommittee that he never questioned Hadley's objectivity in handling the investigation. He was seeking the telephone logs only to be prepared in case NIH was accused of lacking objectivity in the Imanishi-Kari investigation.

Raub believed that a "cryptic remark" that he made might have inadvertently caused Healy to be concerned about Hadley's relationship with O'Toole. "What I meant was as a compliment—that I was impressed that, despite the feelings Dr. O'Toole evinced with respect to the institutional authorities, Dr. Hadley had managed to win her trust and her active involvement in the case," Raub later told Dingell's subcommittee. "I viewed that as a mark of success on Dr. Hadley's part in gaining the full cooperation of Dr. O'Toole."

Healy did not know until a meeting about the telephone logs that Hadley was conducting an authorized investigation into the Tufts and MIT inquiries. That ongoing investigation could explain the telephone calls from O'Toole.

On June 14, in the midst of the fighting over the telephone logs and the Gallo rewrite, *The New York Times* reported on the Storb controversy, quoting the University of Chicago immunologist as refusing to resign. Three days later, Hallum reversed his position and decided to keep Storb on the panel.

Healy also raised questions at the NIH about Hadley's continued stewardship of the Imanishi-Kari and Gallo cases while she was working out of a different office. Supervision over Hadley's work was inadequate, Healy thought. She believed the setup was responsible, at least in part, for leaks about the investigations.

Hallum ordered Hadley to return the case files to the OSI offices and told her that he had been ordered to "rein you in."

On July 1, 1991, Hadley resigned from the two misconduct cases and took a leave of absence.

■

When Hadley returned to NIH in September, she went to work for the Office of Science Policy, where she had what she considered a nothing job developing science education projects. "I was crying all the time," Hadley recalled. Margot O'Toole, the whistle-blower whom the OSI draft report called heroic, called and commiserated with her. "I can't talk about this," Hadley said.

Drawing on her own experience as a whistle-blower, O'Toole warned Hadley: "This is only going to get worse."

"How could it?" Hadley asked.

O'Toole sent her a copy of the Yeats poem that her mother had given her after she was so discouraged by the 1986 meeting with Baltimore, Imanishi-Kari, and Eisen. Hadley put the poem up by her desk at home where she could see it when things, indeed, became worse.

After Hadley dropped her work on the two controversial cases, Healy stepped up her criticism of the way the NIH was investigating scientific-misconduct cases. She suggested a restructuring that would split OSI functions so that one office investigated and another made the judgments of misconduct. She launched an administrative review of the office.

Dingell scheduled a hearing of his oversight subcommittee for August 1 to look into Healy's actions. His staff interviewed Healy on July 19 in preparation for the hearing; Dingell's aides suggested that the continuing case involving the Cleveland Clinic Foundation might be influencing Healy to attack OSI. "However preposterous that was," Healy indignantly told Dingell later, "I removed myself immediately from the chain of command of OSI until the case was resolved."

True to form, prior to the hearing, reports began to circulate in the press about the Cleveland Clinic case and the possibility that Hadley's difficulties with Healy were a result of that investigation. "Congressional investigators are questioning whether the OSI's unprecedented criticism of the NIH director is related to the abrupt withdrawal earlier this month of a key OSI investigator from two highly publicized cases of possible scientific misconduct," wrote reporter John Crewdson in the July 28 issue of the *Chicago Tribune*. Crewdson's remarkably comprehensive reporting on Gallo had sparked the federal investigation into that case.

Dingell launched into an attack on Healy at the hearing, blaming her for setting about to destroy the OSI. He said that "signs of progress" in the NIH's handling of misconduct charges in the Baltimore and Gallo cases "have been virtually obliterated in the past six weeks by unprecedented actions by the new director of NIH, Dr. Bernadine Healy. These actions have derailed two critical investigations . . . demoralized and emasculated the OSI, and made a mockery of OSI's alleged independence in dealing with misconduct allegations," Dingell charged.

Healy's recusal from OSI matters came too late, Dingell said. "The damage had already been done."

Norman Lent, who had attacked Margot O'Toole at the 1989 hearing, was now joining Dingell in blasting Healy and praising OSI and Hadley for their "tenacity, integrity, and fairness." He said the subcommittee staff had "apparently uncovered startling, perhaps even bizarre, behavior by the director of the NIH." Lent, however, was also angry over Healy's approval of a survey of children's sexual behavior, which had gotten the new NIH director in trouble with conservative Republicans right away.

Hadley was called to testify; she spoke of her difficult relations with the new NIH director and the problems she ran into with the Gallo report, the decision to ask for Storb's resignation, and her telephone logs.

■

Healy, like Baltimore, was a worthy opponent for the oversight subcommittee. She was unusually combative in public for a federal bureaucrat facing members of Congress who could make life even more difficult for her agency; her attitude cheered some members of the scientific community upset with Dingell.

She reminded Dingell that when she unexpectedly ran into him shortly after taking office he responded to her request for advice by telling her to follow the rules and procedures. And that she had done, she contended.

Healy defended her handling of the misconduct allegations while she was at the Cleveland Clinic, saying that she had not realized at first that the accused scientist's collaborator was on the original panel. His inclusion was one of the factors that led to her calling for a second inquiry, in which she did not participate. "The clinic took very seriously the allegations and faithfully conducted a multi-level review of this case in compliance with PHS [Public

Health Service] guidelines. . . . I would challenge this committee to ask, 'Where is the coverup?' " she said.

Healy also told of a situation in which OSI officials were confused about their procedures and policies, sensitive papers were kept at a quasi-independent satellite office (where Hadley worked), and lax procedures might have led to breaches in confidentiality that hurt not only the people being investigated but also the investigations themselves. In one instance, a tape of OSI officials talking about how they believed a particular scientist had lied to them and was guilty of misconduct was sent mistakenly to the researcher in question.

After a series of questions about why, as head of research at the Cleveland Clinic, Healy had signed off on the questioned applications for NIH funding, Lent became exasperated with her contention that her approval simply indicated that the institute would support the research, and did not certify the truthfulness of the researcher's scientific statements.

"Mr. Chairman, I can't handle this witness," Lent said at one point. "I am getting unresponsive replies to my questions." Dingell took over, asking about the process of signing the application and why Healy would sign it before the researcher did. "I am just a poor, foolish lawyer from Detroit and I get a little befuddled in some of these difficult questions," Dingell said.

But, strong-willed as Dingell, Healy matched him: "I am just a poor girl from New York."

The hearing ended inconclusively, and Healy would later dismiss it as "good theater" that had "taken on mythic proportions." In an interview, she said that she had since sat down with Dingell and quietly discussed their hopes for NIH; he assured her, she said, that there was no vendetta against her.

■

After Hadley returned to the NIH that fall, she was asked for help by the subcommittee and the inspector general's office, which was investigating Imanishi-Kari for the federal prosecutor in Maryland. This assistance led to further confrontations with NIH officials.

The subcommittee asked Hadley to attend a November 4 meeting at which OSI officials were questioned about whether they were dropping the investigation of Tufts and MIT and also about other issues. NIH officials began accusing Hadley of disclosing confidential information to the subcommittee; they asked her about the panel.

Dingell wrote Healy in December requesting that Hadley be detailed to his subcommittee on an as-needed basis, and again in January. He was turned down by Healy in February.

Hadley, feeling "profoundly ill" from the pressure of all the controversy, went on sick leave in March. On March 11, 1992, the subcommittee's staff asked her to come down to its Capitol Hill office, where she was later tracked down by the NIH and informed that she had to attend to an urgent task at her NIH office. But when she got there, there was no task. Hadley called the personnel office for information on how to resign. While she was waiting for the papers, an FBI agent came into her office. The agent was there at the request of the NIH to investigate the leaking of confidential documents relating to the Gallo and Baltimore cases. Hadley declined to talk before speaking to her attorney. The FBI agent, she said, began telling her how the OSI was being damaged by "documents walking out" of the office and going to the Dingell subcommittee and media.

But, Hadley said, when he asked "Tell me, Dr. Hadley, when did you cease to have lawful access to OSI documents?" she told him to leave.

The next day, Hadley was at the subcommittee offices again; when she called in for her messages at the NIH, a colleague told her that a locksmith had changed the lock on her office door. Hadley never returned. Dingell wrote to Healy protesting Hadley's treatment, contending that the FBI investigation was an "apparent act of harassment and intimidation aimed at courageous, public-spirited whistle-blowers. . . .

"I find it exceedingly curious that the NIH is calling upon the FBI to investigate Dr. Hadley, who has consistently and at great personal sacrifice cooperated with congressional and law enforcement investigations," Dingell wrote.

The following month, in response to a notice from the NIH that her office would be checked for unauthorized confidential documents, Hadley's lawyer and Ned Feder went to her office to pick up her personal belongings. Among the things they brought back was a plaque from the inspector general of the Department of Health and Human Services. It represented an award for her integrity.

The investigation eventually was quietly dropped.

In early July, Dingell wrote to HHS secretary Louis Sullivan requesting that Hadley be detailed to his subcommittee. Sullivan approved the request, allowing Hadley to remain on the NIH payroll and work with the oversight subcommittee on its investigations into how the NIH and the nation's research institutions were handling scientific-misconduct cases.

■

While the NIH was struggling with its efforts to deal with scientific misconduct, the grand jury in Baltimore was hearing witnesses about the *Cell* paper.

It was not until the spring of 1992, after she had first responded to the OSI draft report, that Imanishi-Kari finally got access to her original documents and the Secret Service tests. Bruce Singal turned the documents over to an expert he had retained to examine them.

The expert was Albert Lyter III. Lyter said in his June 17 affidavit that one of the Treasury employees he had helped train was Larry Stewart, who had testified that some of Imanishi-Kari's data appeared to have been cobbled together from experiments other than the ones claimed.

Lyter had been a forensic chemist at Treasury's Bureau of Alcohol, Tobacco and Firearms (ATF) before the responsibility for forensic work on ink samples was transferred to the Secret Service. Since leaving ATF in 1981, Lyter was president of his own consulting company, Federal Forensic Associates Inc. He held a master's in forensic science from George Washington University and was a doctoral candidate in analytical chemistry at the University of North Carolina at Chapel Hill.

According to his affidavit, provided to the U.S. Attorney's Office, Lyter reviewed the thin-layer chromatography (TLC) analysis of the ribbon ink on the radiation-counter tapes in Imanishi-Kari's notebook with the ink on tapes from the other notebooks taken from scientists at MIT. TLC analysis involves placing a speck of the questioned ink on a plastic plate with a thin layer of gel. The sheet is then placed in liquid and the various components of the ink separate out; they then can be analyzed for comparison purposes.

"There are a number of serious flaws which permeate the

Secret Service reports on TLC analysis, and which undermine their conclusions," Lyter said.

He argued that Stewart and Stewart's boss, John Hargett, had taken inadequate samples of data, erroneously interpreted the test results, improperly compared tapes generated by different machines, and had used poor procedures in testing the documents. Lyter contended that there actually was "no chromatographic basis for concluding that the tapes which are contained in the I-1 notebook 1985 were actually generated from the gamma counter on any date earlier or later than the date assigned them."

In order to conclude, as the Secret Service had, that some of the counter tapes were "most consistent with" 1981–1984 tapes from Charles Maplethorpe's notebook, Lyter said said it was necessary to show that Imanishi-Kari's 1985 tapes did not match tapes from the same counter during the same time, and that they do match tapes from the same counter from the 1981–1984 time frame. He argued that the sample of tapes from both time periods was inadequate to draw any conclusions.

The limited sample also made a comparison by paper color useless, he argued, because there were various shades of yellow and green.

Lyter contended that the conclusion about those 1985 tapes was also insignificant because the phrase "consistent with" is "a relatively weak" one, especially compared to one in which the inks "match."

"The phrase 'consistent with,' which is used by the Secret Service, indicates there is much less evidence than what would be necessary to find such a match," Lyter said, also dismissing the qualification of the phrase with "most."

"The Secret Service conclusion that certain of Dr. Imanishi-Kari's 1985 tapes are 'most consistent with' certain of the C-2 tapes furnishes no forensic basis for concluding that the two sets of tapes were generated at or about the same time," Lyter said.

Lyter also argued that Imanishi-Kari's 1985 tapes could not be compared with the Maplethorpe tapes because the former came from a Beckman gamma counter and had a different format than the latter, which came from a Packard beta counter.

In addition, he questioned why the Secret Service did not compare the Imanishi-Kari tapes with some from Mark Pasternak's notebooks dated in 1982, which also apparently came from the same gamma counter. (Pasternak worked in Eisen's lab at the time.) A TLC comparison, he said, showed that her tapes did not match March and April 1982 Pasternak tapes; this apparently indicated that Imanishi-Kari's tapes could not have originated at that earlier time. (Which was not to say, however, that the tapes could not have been dated between April 1982 and the purported 1985 date.)

Lyter also swore that he had discovered errors in the Secret Service findings. In one case, a Secret Service determination that Imanishi-Kari tapes were most consistent with earlier Maplethorpe tapes was based on the presence of a particular type of ink that the agents did not find elsewhere. However, Lyter said the Secret Service, indeed, found that ink in 1985 and 1986 tapes in other notebooks.

Finally, Lyter contended that Stewart and Hargett had conducted the ink analysis on two types of plates of different quality, which improperly added another factor to the comparisons. The tests should have been done on the same type of plate, he said.

(Lyter had gone up against a former colleague before. In fact, he would do so again after conducting the tests for Imanishi-Kari. The Baton Rouge *Advocate* reported on July 31, 1992, that federal forensic scientist Antonio Cantu, a former co-worker of Lyter's, criticized his test results in a case involving alleged false statements to the Internal Revenue Service. Cantu charged that Lyter's ink tests were "meaningless": He did not take multiple samples and he did not thoroughly stir the mixtures used in the tests.)

■

Secret Service agents John Hargett and Larry Stewart were angry when they heard of Lyter's report. In a private briefing for Dingell's staff, they countered Lyter's challenges. In public, they hesitated to say much, but in the great tight-lipped tradition of law-enforcement agencies, they managed to score some points.

They noted that Stewart had chaired a committee the year before to develop standards for comparing inks for the American Society for Testing and Materials. The standards were published in September 1992. Stewart drily added that Lyter had once asked him to go into private business with him, but Stewart had declined.

Lyter's discussion of the use of "consistent with" was playing with semantics, the younger agent contended. "I'm using standard forensic terminology, not weasel words," he said. "I stand behind what we did. Our results were quite strong."

They added that while they initially tested the inks on two different types of plates, they had done so only to determine which one would best discriminate between the inks. They then used only one type of plate for the tests.

Stewart and Hargett also said that while they supported their own ink analysis, their conclusions did not depend solely on those results. In fact, they believed they could throw away the ink analysis and reach the same conclusions based on a combination of other tests and observations, including the great advance in counter numbers on the tapes. That advance showed that either Imanishi-Kari did more than twenty-seven thousand experiments in one day while other scientists did two hundred and thirty, or the tape was out of order, Stewart said.

But Stewart and Hargett never got the opportunity to argue their evidence in court. On July 13, 1992, U.S. Attorney Richard D. Bennett declined to prosecute Imanishi-Kari.

Although he said the Secret Service testimony was persuasive, Bennett did not believe he could persuade a jury beyond a reasonable doubt.

"Even after gaining an understanding of the science, the claim is one about which even reasonable scientific minds can differ, as evidenced by the conclusion of the Office of Scientific Integrity's first review panel," Bennett said in a letter to John Dingell. The Secret Service testimony would be countered by the defense's expert and the statistical analysis would be inadmissible, he said.

Bennett apparently did not agree with the OSI, which in its draft report said that valid results were beside the point if the experiments were fabricated. In his letter, Bennett wrote, "No matter how this case would be charged, the central issue would remain the fundamental validity of Dr. Imanishi-Kari's scientific work. That issue has been debated by scientists for years. If Dr. Imanishi-Kari backdated accurate information in order to organize her sloppy record-keeping but the underlying results were arguably valid, we would be hard-pressed to prove that the backdating was materially misleading," Bennett wrote.

In a letter to Hargett's and Stewart's superior, Bennett stressed that his decision "should not be interpreted as a lack of confidence in them or their testimony. Their testimony would have stood up well against that of the defense expert whose affidavit we received recently."

Assistant U.S. Attorney Geoffrey Garinther, who conducted the investigation, agreed that the Secret Service evidence was "very persuasive," but that "it is possible both to believe the Secret Service and still believe you do not have necessary elements to convince a jury. You must prove corrupt intent."

Supporters of the *Cell* authors claimed victory. Imanishi-Kari's lawyer cheered the prosecutor's decision, contending that the "scandal here is that the investigating agencies have concealed evidence from the accused for several years and not given her the opportunity to defend herself until now."

Baltimore, no longer president of Rockefeller University, said the prosecutor's decision was a vindication and he would retract his retraction. "I think they should apologize for putting [Imanishi-Kari] through six years of hell; they should give her grants back and let her go on being the good scientist she is. I consider that the issue of the scientific validity of this paper is closed."

Dingell, however, was not impressed. Said he: "The decision not to prosecute does not change the fact that the *Cell* paper was retracted because of serious and extensive irregularities."

The prosecutor's decision also did not change the opinion of the so-called Harvard Mafia and others who were critical of the paper and of Baltimore's role in handling the dispute.

"I've had no change of heart," insisted Harvard's John Edsall.

"Scientific matters shouldn't be settled in court. That's just a legalism," added Serge Lang, the Yale mathematician whose papers and clippings on the case influenced Edsall to take another look at the controversy.

Harvard's Walter Gilbert said he believed there still was a strong case to be made for fraud despite Lyter's affidavit and the prosecutor's decision not to indict. However, he added, "this case has less to do with fraud and much more to do with notions of correct behavior."

NINETEEN

The Aftermath

■

Throughout the twentieth century, scientists have been portrayed as heroes, mythic figures like Marie Curie, Albert Einstein, and Robert Oppenheimer, people more noble than the rest of us, and certainly smarter. Sometimes eccentric, but always seekers of truth, scientists have served as the human symbols of the modern faith in technology. The central lesson of the Baltimore–Imanishi-Kari controversy is that scientists are not necessarily noble—and, more important, for them as for all of us, truth can be agonizingly elusive.

It is fairly well accepted by many of his friends and critics that Baltimore's story has elements of a classic Greek tragedy. If Baltimore was guilty of anything it was hubris, the sin of pride. He truly could have stopped the dispute at many different points—when O'Toole first approached MIT; after the 1988 hearing; and before or at the 1989 hearing—by simply saying "Yes, there appear to be serious questions here. We're going to withdraw the paper until we can sort them out." The damage to his reputation would have been nil, and David Weaver certainly could have been protected by Baltimore's patronage. But Baltimore's arrogance would not let him back off, and because of it he was eventually forced to resign the presidency of Rockefeller University.

This story also would not have been the same had Margot O'Toole not been so sure of herself, so certain that she was right, and so unable to let the facts slide, even after she was vindicated by the OSI draft report. The zeal for pursuing misconduct also propelled Walter Stewart and Ned Feder, who were unconcerned that other scientists believed that they were wasting their time at "destructive" work rather than doing "real" science.

Another key to this story's going as far as it did was John Dingell's doggedness—his willingness to employ every weapon he had, including a well-honed ability to use the news media, and his lack of fear of bad press. He wasn't fair to Imanishi-Kari and Baltimore, or to the MIT and Tufts reviewers, in 1988, when he did not give them a chance to respond to the clear implication that he believed in O'Toole. Dingell defended himself against such charges of unfairness by contending that an inquiry typically first collects the allegations that there are problems. But his second congressional hearing, in which Baltimore and the others could respond, was not held for another year. However, without Dingell, the NIH would never have reopened its investigation—and, having opened it, might not have had the will to see it through, had Dingell not maintained his pressure.

As for Thereza Imanishi-Kari: Her loyal following was impressive; some people just could not believe she was guilty of anything more than sloppiness, a temper, and dedication to the lab. I, however, found the Secret Service findings and the draft OSI report more persuasive.

Some friends and critics suggested that the problems with the paper came at a time when she was suffering great pain from lupus. Imanishi-Kari has not used that as an excuse; she insists that she did what she published, only she can't prove it. O'Toole is oddly sympathetic toward Imanishi-Kari, believing that she was willing to correct the paper but was told that it was unnecessary to do so. O'Toole says Imanishi-Kari is also a victim of the scientific establishment.

No matter what happened, the controversy seems never to end. This was no less true when the federal prosecutor decided against indicting Thereza.

Imanishi-Kari's friends Joan Press of Brandeis and David Parker of the University of Massachusetts began a campaign to stir scientists to write to the new Office of Research Integrity and to NIH officials to clear the Tufts scientist and restore her funding. Tufts maintained Imanishi-Kari's salary, but federal grants are the lifeblood of scientists; they provide crucial support for laboratories and research assistants. Imanishi-Kari's name was taken off the alert system pending a final outcome of the government's investigation, but as of mid-1993 she was working without a federal grant.

"It is crucial that the ORI and the NIH know that there is continuing strong concern in the scientific community about how Imanishi-Kari has been treated, and that we will not ignore further misconduct or unfair treatment in this investigation," Press and Park said in an October 5, 1992, letter to scientists around the country.

The Office of Research Integrity had taken over the case from the OSI and Office of Scientific Integrity Review, which were abolished three years after being established to preempt legislative solutions that were floating in Congress. The responsibility for handling scientific misconduct was removed from the NIH; the ORI is part of the Department of Health and Human Services. Once the new office investigates allegations and finds scientific misconduct, the accused may take the case before an open, formal hearing by the departmental appeals board and an administrative law judge. The final government regulations providing for the formal appeals hearings, updating the rules governing how research institutions must deal with misconduct allegations, and defining scientific misconduct were expected to be reviewed by the new Clinton administration before being published.

Ironically, scientists who believed that they could best police and judge themselves were moving toward a system more dominated by lawyers, largely as a consequence of their complaints that Imanishi-Kari and other accused researchers were not getting due process from the NIH. Even James Wyngaarden, the former NIH director who fought to establish the OSI within his agency in order to avoid a takeover by lawyers, now believes that the original system was not fair enough to scientists charged with wrongdoing.

Bernadine Healy, who called OSI a "star chamber," says it was just as well that investigations of scientific misconduct were taken out of NIH. Scientists, she said in an interview, do not have the "mindset" to believe that other scientists could lie.

Breckinridge Willcox, the U.S. attorney during the early stages of the Imanishi-Kari investigation, contended that NIH itself shared the blame for "the scientific community's refusal to deal forthrightly with the problem," because of the agency's reluctance to pursue cases of suspected fraud. Willcox has argued powerfully in favor of selective criminal enforcement, contending that scientists "motivated by economic or self-aggrandizing considerations" should be treated no differently from other professionals, who are "routinely prosecuted for providing false information to the government."

Clearly, if public and congressional watchdogs like John Dingell do not begin to believe again that the scientific establishment has the will to police itself, they will look increasingly to criminal prosecution for dealing with allegations of scientific misconduct.

There are signs that the new investigative office established within HHS is going to step up to the plate. On December 30, 1992, the Office of Research Integrity found that Robert Gallo had committed scientific misconduct and intentionally misled colleagues about whether he had grown or studied a French strain of the AIDS virus. The ORI's report, which was leaked to the press, reversed earlier findings by Bernadine Healy; it said that Gallo had grown the French virus and used it in his own research. The report came almost a year and a half after Healy's pressure forced Suzanne Hadley off the Gallo and Baltimore investigations.

The new HHS office also was expected to generally uphold the OSI's draft findings on the Imanishi-Kari case, according to sources who talked to federal officials involved in the inquiry. Imanishi-Kari could then appeal to the departmental board in the fight for her reputation and funding. She has since published results of new experiments that she said confirm much of her earlier work. Another team of scientists, including Stanford's Leonore Herzenberg and Columbia's Alan Stall, confirmed aspects, as well; the team's findings were scheduled for publication in the fall of 1993. But as the earlier OSI draft report noted, possibly accurate results would not excuse fraudulent experimental work: It doesn't matter if the answers are right if you make up the experiments.

Although the NIH and HHS initially were able to avoid congressional action establishing a misconduct office, lawmakers finally codified the Office of Research Integrity as an entity in HHS and called for the formation of a commission to make recommendations on the establishment of such an office. The legislation, signed into law by President Bill Clinton on June 10, 1993, also was designed to afford more protection to whistle-blowers. The measure originally had been developed by Dingell and the late Ted Weiss and had been approved by Congress as part of a larger 1992 bill that was vetoed by President George Bush because it would have allowed fetal-tissue research.

But if scientists believed that Dingell's interest would end there, they were sadly mistaken. As of summer 1993, John Dingell's subcommittee aides were still looking into the responses by MIT and Tufts to Margot O'Toole's questions about the *Cell* paper and considering whether there had indeed been a cover-up. Suzanne Hadley has stayed on the NIH payroll but worked on the MIT-Tufts investigation for the subcommittee, as well as on the panel's own inquiry into whether Robert Gallo misappropriated HIV from French scientists.

Margot O'Toole works on cancer research at the biotech company founded by Mark Ptashne. She was awarded the Cavallo Foundation prize for moral courage and was nominated by John Edsall and Paul Doty for another prize from a professional society. O'Toole told friends that the $10,000 Cavallo prize would help pay old debts from the time when she was out of the lab.

Her thinking about right and wrong remains clear as ever. At a Washington, D.C., reception honoring the Cavallo Award–winners, O'Toole said she had been astonished to run up against an "us versus them" mentality. "The 'us' are the scientists who are supported by public funds and the 'them' are the public. I was told that we sometimes have to hide our mistakes, cut corners, stretch the truth, and embellish our accomplishments, or else 'they' might stop giving money to us," she said. "What I did was nothing more than what I believe was the minimal requirement for someone wishing to call herself a scientist."

But in a compassionate way, O'Toole also warned about the ease with which people, in the pursuit of their life's work, can lose sight of their ideals and their commitment to the truth. "By the time we are expert enough to spot something wrong, we have become insiders. We have spent our lives working to become part of the group, finding a niche suited to talents and aspirations. . . ." she said. "It is very difficult to take actions which could adversely affect one's friends and colleagues."

Walter Stewart and Ned Feder have lost none of their drive to root out misconduct. Still at the NIH, they have developed a computer program to catch plagiarism in scientific reports. They have also begun to study the experiments of a prominent scientist, whom Stewart declined to identify, whose work has been challenged. Their telephone constantly rings with scientists complaining about the shoddy work of others.

Although Stewart holds fast to what he calls "kindergarten eth-

not so much fraud as "the stifling, authoritative atmosphere" of mainstream scientific research that pressures people to publish and compete for funds. "I'm interested in fraud because of the nexus to the lousy environment we're working in. We've got to create an environment that allows people to be creative. . . . What we need to do is not [only] catch the cheaters but change how we do science."

In April 1993, NIH split up Stewart and Feder and assigned them other tasks. Believing the decision was in retaliation for their fraud work, they fought back. Stewart, fearing that other whistle-blowers would be neglected, went on a hunger strike for a month until Dingell aides agreed to look into his concerns.

By the end of 1992, David Baltimore was preparing to leave New York and return to Cambridge. If he wasn't going to be able to put his administrative abilities to use at rebuilding Rockefeller University, there was little reason to stay. He believed MIT was a better place to do science.

Asked what he might have done wrong in the whole case, Baltimore sounds very much like Imanishi-Kari, who believes her major error was in not shaking O'Toole's proffered hand at the end of the meeting with Wortis and Huber.

"I should have taken the threat of Stewart and Feder more seriously," Baltimore told me, reflecting on the dispute, then six years old. Baltimore believes the other major error was Imanishi-Kari's: She should, he said, have sent her data papers unorganized to Dingell's subcommittee.

Having been taken to task by his colleagues at Rockefeller and elsewhere for backtracking, Baltimore walked a fine line in his statements about the *Cell* paper and O'Toole. Although he initially said he would seek to retract his retraction, he told me that he was delaying that action to let the whole dispute run its full course at HHS.

He often fell off that line, however, when he discussed the events with me. He wasn't retracting his apology to O'Toole, he said, but he did contend that she "selected" the seventeen pages that would prove her points. "Some of my best friends are wrong about things. I don't think she's being honest to herself," he said.

■

Scientists around the country were torn by the controversy and found themselves unable to decide clearly who was right and who was wrong. Many remained angry with John Dingell, Walter Stewart, and Ned Feder for what they considered undue government intrusion into scientific issues, while at the same time they agreed that if Imanishi-Kari was not guilty of making up data, she was at least so sloppy that her work on the *Cell* paper was unacceptable and precluded any clear proof of her innocence. Baltimore was guilty of terminal arrogance that led his colleagues—guilty themselves of accepting his word rather than looking at the facts of the case—into a hapless confrontation with Congress and a loss of prestige in the public mind.

One prominent scientist, who redid the work of an assistant when it was challenged by another researcher, still feels that the scientific and academic communities are under assault from a growing anti-intellectual movement, an assault that started with Ronald Reagan's Republican administrations and includes Dingell's investigations and the growing politicization of the NIH, in which merit is not always a top priority in grantmaking. "There's a broad-based view that these pointy heads are more trouble than they're worth. Nobody wants to pay the cost of anything," he says. "In this environment, which was an attack on the citadel, [Baltimore] responded in an arrogant way and brought this down on himself. I fault David in his loss of perspective."

Baltimore was supposed to be better than that. He was supposed to be more savvy than to stonewall Dingell, whose political agenda might have included attacking the big bad guys but who certainly could not have any interest in scarring the scientific community. Baltimore was supposed to be secure enough and scientifically dispassionate enough to hear O'Toole's challenge and tell his collaborators that the work had to be thoroughly reviewed and possibly redone. To many of his friends, and friends-turned-critics, Baltimore's behavior was inexplicable.

In the flurry of meetings, debates, and conferences on scientific ethics that were held in the aftermath of the Baltimore and Gallo cases as well as revelations of scientific misconduct elsewhere, scientists wondered what could or should be done to counter the wrongdoing and the growing public perception that scientists held no higher moral ground than bankers or car dealers. Research institutions that did not have guidelines drafted them; others tried to get the word out, particularly to young scientists, that sloppy work and misrepresentation—to say nothing of outright fraud—were not acceptable.

Some scientists question whether there is in fact a significant, growing problem with scientific misconduct, or simply a public-relations difficulty. The scientific community has not been able to get a handle on the problem, although most scientists would admit that one fraud is one too many. At a time when some $9 billion in NIH grants is at stake and there are more scientists than ever competing for the funds, the temptation to shave one's results is great.

Many of the problems seem to result from the old-boy networks and the self-interest that permeates the scientific community and decides who gets funded, who gets credit, who gets published, who gets promoted.

Peer review clearly still has its place; one would hate to see decisions on the funding of scientific work left solely to congressional committees with political interests, or to closeted bureaucrats. One of the lessons of the Baltimore–Imanishi-Kari case seemed to be, as Paul Doty believed, that established scientists must take a public stand on just what is acceptable scientific behavior.

Stewart and Feder mourn the lack of other moral leaders of the stature of John Edsall and Paul Doty or the late Richard Feynman. Stewart and Feder are not alone; James Watson, the great DNA scientist, spoke of that problem in a 1990 interview with *Science & Government Report.* Asked about the Baltimore case, Watson said: "One of the problems now is that because there are so many scientists, and it's so big, there are no perceived moral leaders in our profession, in the sense that Niels Bohr was perceived in physics. Someone who sets a standard for behavior. People need role models."

Howard Temin, who shared the Nobel Prize many years ago with David Baltimore, was reluctant to discuss the *Cell* controversy in detail. His take on it, however, also speaks to the moral leadership that is required of all scientists.

"If I could make a rule it's this: Science is of data and experiments and not of people and authority," Temin said. "It is the responsibility of the people who have done the experiment to look at the data without considering the standing of who is raising the questions. David is one of the most successful and productive molecular biologists in the world. But it doesn't matter who you are. Science is the facts and the techniques."

Notes

■

Unless otherwise specified, all interviews were conducted by the author.

Chapter One

Page 1. . . . *revolutionary discovery was called idiotypic mimicry* . . . "Altered Repertoire of Endogenous Immunoglobulin Gene Expression in Transgenic Mice Containing a Rearranged Mu Heavy Chain Gene," by David Weaver, Moema H. Reis, Christopher Albanese, Frank Costantini, David Baltimore, and Thereza Imanishi-Kari, *Cell*, April 25, 1986. Also see "Trouble in the Laboratory," by Kathy A. Fackelmann, *Science News*, March 31, 1990, page 200: and "Deciphering the Science," by David P. Hamilton, *Science*, March 8, 1991, page 1170.

Page 2. *Margot was the daughter* . . . Interviews with Margot O'Toole and Elizabeth O'Toole.

Page 2. *Margot's life path* . . . Margot O'Toole interview.

Page 3. . . . *with the help of Henry Wortis* . . . Interviews with Margot O'Toole and Thereza Imanishi-Kari: testimony of Henry Wortis, May 9, 1989, House Energy and Commerce Subcommittee on Oversight and Investigations.

Page 4. . . . *one source of tension* . . . Interviews with Brigette Huber and O'Toole. Also see "The Mind of a Whistle Blower," by Barbara Carton, *The Boston Globe*, April 1, 1991; "Researcher Blew the Whistle Once Before," by Judy Foreman, *The Boston Globe*, May 10, 1989, page 3.

Page 4. *But Margot was unable to duplicate* . . . O'Toole and Imanishi-Kari interviews.

Page 5. *Now out on his own* . . . Interview with Charles Maplethorpe, and congressional testimony, Oversight and Investigations Subcommittee, April 12, 1988.

Page 5. *Finally, Margot acknowledged failure* . . . O'Toole and Imanishi-Kari interviews; Wortis congressional testimony; transcript of June 4, 1991, meeting of Herman Eisen and Harvard scientists.

Page 7. . . . *forces that she had unleashed* . . . Interviews with O'Toole, Huber, and Robert Woodland.

Page 8. *O'Toole also tracked down Martin Flax* . . . Martin Flax and Wortis congressional testimony of May 9, 1989.

Page 9. *Wortis, Huber, and Woodland met* . . . Huber and Woodland interviews; Wortis congressional testimony.

Page 10. *But the memories of the participants* . . . Interviews with O'Toole, Imanishi-Kari, Huber, and Woodland.

Page 10. *In a letter written a year later* . . . Wortis letter of June 17, 1987, to Henry Banks.

Page 10. *Flax also asked Dr. Sidney Leskowitz* . . . Flax congressional testimony, May 9, 1989.

Page 11. . . . *voted to recommend that Imanishi-Kari be appointed.* Interviews with O'Toole, Peter Brodeur, and aides to Representative John Dingell; Flax congressional testimony.

Page 11. *While she remained opposed to filing formal charges* . . . O'Toole interview.

Page 12. *Brown, however, recalled a collegial conversation* . . . Gene Brown congressional testimony, May 9, 1989.

Page 15. *The next day* . . . Eisen memo of June 17, 1986.

Page 17. *They were unable to afford* . . . O'Toole interview.

Page 18. . . . *give her daughter the courage* . . . Elizabeth O'Toole interview.

Page 18. . . . *Despite his earlier warning* . . . Maplethorpe interview.

Chapter Two

Page 20. *A clue may be found* . . . *Immunology*, by Richard M. Hyde, Harwal Publishing Company, 1992, pages 35–43, 88; "Deciphering the Science," by David P. Hamilton, *Science*, March 8, 1991, page 1170.

Page 21. *Jerne won the Nobel* . . . "Three Immunology Investigators Win Nobel Prize in Medicine," by Harold M. Schmeck Jr., *The New York Times,* Oct. 16, 1984.

Page 21. *". . . I am not the oddball . . ."* "Nobel Winners: Three Quiet Men Conquering Hidden Mysteries," by John Noble Wilford, *The New York Times,* Oct. 16, 1984.

Page 21. . . . Cell *paper* . . . Weaver et al., *Cell,* 1986.

Page 22. . . . *Baltimore wanted to take advantage* . . . "Baltimore's Travels," by David Baltimore. *Issues in Science and Technology*, Summer 1989, page 48.

Page 22. *"Our motivation was . . ."* Interview with Frank Costantini.

Page 22. *As Baltimore would later tell* . . . May 19, 1988. Baltimore letter to scientists around the nation.

Page 23. *"Studies of the expression . . ."* "Baltimore's Travels," *Issues in Science and Technology*, page 48.

Page 23. . . . *others saw hints that the results* . . . "Trouble in the Laboratory," *Science News,* March 31, 1990; "Deciphering the Science," *Science*, March 8, 1991, page 1170; "The Transgene Factor," by Frank Kuznick, *The Washington Post Magazine*, April 14, 1991, page 26.

Page 24. *". . . very frequently mimics . . ." Cell*, April 25, 1986.

Page 26. *In fact, Baltimore says now* . . . Interview with David Baltimore.

Page 27. *Imanishi-Kari would later say* . . . Interview with Imanishi-Kari.

Chapter Three

Page 28. *The Harvard-Emory fraud* . . . Interviews with Walter Stewart and Ned Feder; testimony before the Oversight and Investigations Subcommittee, April 12, 1988.

Page 30. . . . *he couldn't encourage Stewart and Feder* . . . "Why We Should Change the Libel Law," by Floyd Abrams. *The New York Times Magazine*, September 29, 1985, page 34.

Page 31. . . . *Batman and Robin*. "Fraud Busters," by Frank Kuznik, *The Washington Post Sunday Magazine*, April 14, 1991, page 22.

Page 32. *Stewart did not shy away* . . . Stewart interview; "Montgomery Council Trims Lawn Law," by Sue Anne Pressley, *The Washington Post*, November 16, 1988, page B3; "Md. Couple Refuses to be Mowed Down," by Jo-Ann Armao, *The Washington Post*, June 16, 1988, page D1.

Page 33. *Maplethorpe called that night* . . . Maplethorpe interview.

Page 33. *Maplethorpe outlined O'Toole's story* . . . Interviews with Maplethorpe, Stewart, and Feder.

Page 34. *Then, at one of their lunches* . . . Interviews with Maplethorpe and O'Toole.

Page 35. *The only thing for him* . . . Interviews with O'Toole, Stewart, and Elizabeth O'Toole.

Page 35. *"I will always tell you . . ."* Interviews with O'Toole and Stewart.

Page 35. *Understanding the data* . . . Interviews with Stewart and Feder.

Page 36. *" . . . These data appear to have a direct bearing . . ."* Stewart and Feder draft.

Page 37. *When Stewart and Fader initially contacted them* . . . Stewart and Feder interviews; notes; Baltimore letter to Stewart of Jan 21, 1987.

Page 38. . . . *retaliation for their campaign* . . . Stewart and Feder letter to scientists on November 21, 1986.

Page 39. *One of the scientists who responded* . . . "Government and the Freedom of Science," by John Edsall, *Science*, April 29, 1955, page 615.

Page 40. *The performance rating* . . . Jan. 6, 1987, memo from Pierre Renault to Feder.

Page 41. *Baltimore proposed that* . . . Baltimore letter to Joseph E. Rall dated March 17, 1987; Baltimore letter to Walter Stewart and Ned Feder dated March 24, 1987.

Page 43. . . . *Stewart and Feder were prohibited* . . . Rall letter to Baltimore of April 2, 1987; Stewart and Feder memo to Joseph E. Rall of April 9, 1987.

Page 44. *Lewin responded fairly promptly* . . . Benjamin Lewin letter to Walter Stewart of October 19, 1987.

Page 45. . . . *Morgan, too, declined to publish*. Patricia Morgan letter to Walter Stewart of December 16, 1987. Also see "Fear of Suits Blocks Retractions," by A. J. Hostetler, *The Scientist*, October 19, 1987, page 1.

Chapter Four

Page 46. . . . *the son of a Polish immigrant* . . . Interviews with John Dingell, James Dingell, and congressional aides. Also, see *Politics in America 1992*. CQ Press, 1991, pp. 766–69; "Rep. Dingell Wields Wide Power to Probe Much of U.S. Industry," by David Rogers, *The Wall Street Journal*, March 5, 1990, page 1; and "Michigan Democrat Presides as Capital's Grand Inquisitor," by David E. Rosenbaum, *The New York Times*, September 30, 1991, page 1.

Page 48. *"I had fallen in love . . ."* James Dingell interview.

Page 49. *Dingell wooed Deborah Insley* . . . "Big John," by Bob Dart, *The Detroit News Magazine*, May 26, 1985, page 11.

Page 49. . . .*"live with ambiguity"*. . . Timothy Wirth quoted in "Powerful Energy Panel Turns on Big John's Axis," by David Maraniss, *The Washington Post*, May 15, 1983.

Page 50. . . . *"crucial decisions are made . . ."* "Powerful Energy Panel," *The Washington Post*.

Page 50. . . . *lobbyist Thomas Tauke* . . . "Powerful Energy Panel," *The Washington Post*.

Page 51. *James Scheuer, once the second-ranking* . . . "Raging Bulls," by W. John Moore, *National Journal*, January 23, 1993, page 192. Also see "Swinging the Big Gavel," by Joyce Barrett, *M*, March 1990, page 127.

Page 52. . . . *there has been an embarrassing misstep* . . . "Lobby Talk," by Judy Sarasohn, *Legal Times*, March 27, 1989, page 5. Also see "Dingell Calls Off Inquiry Into Investigator's Role," by Kurt Eichenwald, *The New York Times*, March 22, 1989, page D2.

Page 53. . . .*"kennel of junkyard dogs"*. . . "Energy and Commerce," *National Journal*, June 15, 1991, page 1430.

Page 53. . . . *members of the "Dingell bar"*. . . "Specializing in Dingell," by Terence Moran, *Legal Times*, May 28, 1990, page 1. Also see "Enduring a Congressional Investigation," by James Fitzpatrick, in *Litigation*, Summer 1992, page 16.

Page 53. *"We kick their butts,"* Interview with a Dingell aide.

Page 54. *"Whenever we're getting . . ."* "Hill Employee Says He Gave Documents to Post to Get Publicity for Hearings," by Kenneth Bredemeier, *The Washington Post*, July 20, 1982, page 13. Also see "Committee Aide's Work for Media Is Challenged," by Jonathan Friendly, *The New York Times*, January 4, 1982, page B8.

Page 54. *"It's the best act . . ."* Stockton interview.

Page 54. *". . . totally illegible handwriting."* Interview with Michael Barrett.

Page 55. . . . *Stockton called Stewart* . . . Interviews with Stockton, Stewart, and Bruce Chafin.

Page 57. . . . *it should not appoint* . . . Interview with Robert Charrow.

Page 58. *At a talk with science writers* . . . Transcript of Baltimore talk with D.C. Science Writers Association, May 2, 1989.

Page 59. "*The hearings were rather a surprise* . . ." "Akin, Gump's Science Project," by Judy Sarasohn, *Legal Times*, July 4, 1988, page 1.

Page 59. . . . *pleased with their main witness* . . . Chafin interview.

Chapter Five

Page 60. *At 10:10 A.M.* . . . Oversight and Investigations Subcommittee, April 12, 1988.

Chapter Six

Page 76. . . . *learned from a newspaper reporter* . . . "Baltimore's Travels," by David Baltimore, *Issues in Science and Technology*, Summer 1989, page 50.

Page 77. *Dr. Phillip Sharp, head of the* . . . Interview with Phillip Sharp.

Page 77. *"He was one of those people . . ."* Interview with Norton Zinder.

Page 77. . . . *at the Jackson Lab* . . . Interview with Howard Temin.

Page 77. *That was a special summer* . . . Baltimore interview.

Page 78. *Baltimore would later explain . . .* Baltimore congressional testimony, May 9, 1989.

Page 79. . . . *Baltimore made plans . . .* Baltimore interview; "Three Share in Nobel Prize for Work on Viruses and Genes" and "Winners of Nobel Prize in Medicine," by Victor K. McElheny, *The New York Times,* October 17, 1975, page 12.

Page 79. *Baltimore tried calling . . .* Baltimore interview.

Page 83. *"We felt he needed . . ."* "Akin, Gump's Science Project," *Legal Times,* July 4, 1988.

Page 83. *"I didn't know . . ."* Baltimore interview.

Page 83. *Baltimore said it was his own idea . . .* Baltimore interview.

Page 84. *Leonore Herzenberg later said . . .* Interview with Leonore Herzenberg.

Page 85. *Stockton, who contended . . .* Stockton interview.

Chapter Seven

Page 88. *The cigarette butts . . .* Imanishi-Kari interview.

Page 90. *She had known and worked . . .* Interviews with Imanishi-Kari and Baltimore.

Page 91. *"Thereza is very intense . . ."* Interview with Joan Press.

Page 91. *"Maybe I do . . ."* Imanishi-Kari interview.

Page 92. *"I think she . . ."* Imanishi-Kari interview.

Page 92. *"The reason [the paper] favors . . ."* Imanishi-Kari interview.

Page 93. *"I'm not asking . . ."* Imanishi-Kari interview.

Page 93. . . . *relations with Margot became more tense . . .* Interviews with Imanishi-Kari and O'Toole.

Page 93. *Thereza had been bone-tired . . .* Imanishi-Kari interview.

Page 94. . . . *she didn't remember crying . . .* Imanishi-Kari interview.

Page 94. *". . . my mistake was . . ."* Imanishi-Kari interview.

Page 95. . . . *another fateful event . . .* Interviews with Imanishi-Kari, Baltimore, and Normand Smith; transcript of NIH consultants panel meeting with *Cell* authors and attorneys; "Baltimore Declares O'Toole Mistaken," by David Baltimore, *Nature,* May 30, 1991, page 342; Office of Scientific Inquiry draft report, March 14, 1991.

Page 95. . . .*"as comprehensible as possible."* As quoted by the Office of Scientific Integrity draft report, March 14, 1991.

Page 95. *"He didn't force me."* Imanishi-Kari interview.

Chapter Eight

Page 96. *"If Baltimore hadn't raised . . ."* Barrett interview.

Page 98. *Hargett laid out . . .* Interviews with John Hargett and Larry Stewart of the Secret Service. Also see Hargett and Stewart congressional testimony May 4, 1989, and May 14, 1990.

Page 99. *Speaking of the Dingell aides . . .* Hargett interview.

Page 103. *However, by exposing . . .* Herzog congressional testimony, May 4, 1989.

Page 103. *Presentation of their findings . . .* Hargett and Stewart interviews.

Chapter Nine

Page 105. *"We needed a preemptive strike"*. . . Interview with James Wyngaarden.

Page 106. *"When people's careers . . ."* Charrow interview.

Page 106. *"Most of these . . ."* Wyngaarden interview.

Page 106. *Disdainful of the . . .* Charrow interview.

Page 108. *Davie, McDevitt, and Storb were highly critical of*. . . Panel report released by the NIH February 1, 1989.

Page 109. *The error in not distinguishing . . .* Congressional testimony by Joseph Davie, May 4, 1989.

Page 109. *In a meeting with Imanishi-Kari . . .* Transcript of NIH panel meeting, May 3, 1989.

Page 110. *"Where the panel is critical . . ."* November 29, 1988, letter from Baltimore, Imanishi-Kari, and David Weaver to M. Janet Newburgh.

Page 111. . . . *O'Toole was unsatisfied.* Letter from O'Toole to Newburgh of November 28, 1988.

Page 112. *Imanishi-Kari was also questioned . . .* As quoted by Katherine Bick, deputy director for extramural research, in January 18, 1989, memo to Wyngaarden.

Page 112. . . . *submit the proposed corrections . . . Cell* (May 19, 1989) published a letter from Imanishi-Kari, Weaver, and Baltimore, sent at the request of the NIH director. The letter included unpublished subcloning data to bolster the original paper.

Page 113. *Imanishi-Kari told one of Wyngaarden's aides . . .* Wyngaarden congressional testimony, May 4, 1989.

Chapter Ten

Page 114. *Scientists had come . . .* "The Dingell Probe Finally Goes Public," by Barbara J. Culliton, *Science*, May 12, 1989, page 646.

Page 114. . . . *a seat up front for Mark Ptashne . . .* Interview with Mark Ptashne.

Page 115. *The January conference . . .* Interviews with scientists at the conference. Also see "A Clash of Cultures at Meeting on Misconduct," by William Booth, *Science,* February 3, 1989, page 598.

Page 116. *Rather than bring . . .* Zinder interview.

Page 119. . . . *if he looked shaken . . .* Baltimore interview.

Page 119. *Baltimore went on the offensive . . .* Baltimore talk to D.C. Science Writers Association, May 2, 1989.

Page 122. . . .*"We have reopened . . ."* Wyngaarden congressional testimony, May 4, 1989.

Page 122. *Bruce Singal . . . started the session . . .* Transcript of NIH panel meeting, May 3, 1989.

Chapter Eleven

Page 128. *Gaveling the hearing . . .* Oversight and Investigations Subcommittee hearing, May 4, 1989.

Page 130. . . . *Russell Smith had brought Baltimore . . .* Interview with Russell Smith.

Page 138. *For the record* . . . One of the papers was "Isotype Switching by a microinjected Mu Immunoglobulin Heavy Chain Gene in Transgenic Mice," by Jeannine Durdik, Rachel M. Gerstein, Satyajit Rath, Paul F. Robbins, Alfred Nisonoff, and Erik Selsing, *Proceedings of the National Academy of Sciences USA*, April 1989, page 2346. Selsing wrote Baltimore in a May 1, 1989, letter that while some of the findings of his group's paper were consistent with the *Cell*, "in contrast to one of your hypothesis [sic], we do not believe that our work provides evidence for idiotypic mimicry."

Page 144. *"It was heartwarming . . ."* "Dingell Probe Finally Goes Public," *Science*, May 12, 1989, page 646.

Chapter Twelve

Page 145. *On the docket were* . . . Oversight and Investigations Subcommittee hearing, May 9, 1989.

Chapter Thirteen

Page 163. *Dingell personally responded* . . . Dingell letter to the editor, *The Detroit News*, May 30, 1989.

Page 165. *A rare sympathetic treatment* . . . "Anatomy of a Scientific Scandal: O'Toole's Whistle," by Diana West, *The Washington Times*, June 15, 1989, page E1.

Page 165. . . . *"an old infantryman"*. . . Interview with John Dingell.

Page 165. . . . *"Harvard Mafia"* . . . Interview with Mark Ptashne.

Page 166. *During a break* . . . Interviews with Ptashne and O'Toole.

Page 167. *Gilbert was a longtime friend* . . . Interviews with Walter Gilbert and Phillip Sharp.

Page 167. *". . . a bad motivation . . ."* Gilbert interview.

Page 169. *Baltimore's name surfaced* . . . "Dispute on New President Shatters Tranquil Study at Rockefeller U.," by William K. Stevens, *The New York Times,* October 10, 1989, page 1; and "Rockefeller U., Still Affected by Charges Involving Its President, Asks Anew Whether Its World-Class Reputation Can Endure," by Debra E. Blum, *The Chronicle of Higher Education*, September 25, 1991, page A25. Also see "David Baltimore's Final Days," by Stephen S. Hall, *Science*, page 1576; and "A Troubled Homecoming," by Tim Beardsley, *Scientific American*, January 1992, page 33.

Page 169. *In the face* . . . "David Baltimore's Final Days," *Science*, December 13, 1991, page 1576.

Page 170. . . . *"in considerable depth"*. . . *The New York Times*, October 10, 1989.

Page 170. *But the faculty members . . . were not placated.* Zinder interview.

Page 170. *Baltimore made things worse* . . . *The New York Times*, October 10, 1989.

Page 170. *". . . bygones should be bygones . . ."* "Baltimore Accepts Post Despite Faculty Outcry," by William K. Stevens, *The New York Times*, October 18, 1989, page B9.

Page 170. *In the midst* . . . "Ruckus at Rockefeller," *The New York Times*, October 29, 1989, page 28.

Page 171. *"For me, coming here . . ." Science*, December 13, 1991, page 1576.

Page 171. *"There are certainly events. . ." The New York Times*, October 18, 1989, page B9.

Chapter Fourteen

Page 177. *Two days later, Hadley informed Imanishi-Kari . . .* January 12, 1990, letter from Hadley to Imanishi-Kari.

Page 179. *"The refusal of . . ."* December 11, 1989, letter from Dingell to Singal.

Page 181. *Singal started out . . .* Transcript of May 10, 1990, news conference.

Page 185. *In Dingell's opening remarks . . .* Oversight and Investigations Subcommittee, May 14, 1990.

Page 188. *On October 13 . . .* Transcript of October 13, 1990, meeting with NIH panel and Office of Scientific Integrity investigators.

Page 189. *A Science & Government Report interview . . .* "Q&A With James Watson, Genome Project Chief," March 15, 1990, page 4.

Page 190. *"I always felt . . ."* Interview with John Edsall of Harvard.

Chapter Fifteen

Page 191. *"It remains unclear . . ."* OSI draft investigative report, March 14, 1991.

Page 194. *Imanishi-Kari would later complain . . .* Imanishi-Kari May 23, 1991, response to draft OSI report.

Page 208. *"Very often, at that time . . ."* "Verdict in Sight in the 'Baltimore Case,' " by David P. Hamilton, *Science*, March 8, 1991, page 1171.

Chapter Sixteen

Page 214. *". . . that sloppiness or fraud . . ."* "The Imanishi-Kari Case: Can the Scales of Justice Be Calibrated for Scientific Fraud?," by Philip J. Hilts, *The New York Times*, March 31, 1991, Section 4, page 7.

Page 215. *A* Time *essay . . .* "Science, Lies and the Ultimate Truth," by Barbara Ehrenreich, *Time*, May 20, 1991, page 66.

Page 215. *In the midst . . .* "MIT Institute Used Funds Wrongly," by Judy Foreman, *The Boston Globe*, April 17, 1991, page 1; "Baltimore Seems to Be Moving to Clear His Name," by Judy Foreman, *The Boston Globe*, April 24, 1991; "M.I.T. Will Pay Back $731,000 to U.S., Including Lobby Funds," by Philip J. Hilts, *The New York Times*, April 24, 1991, page 16. Also see "Akin, Gump's Science Project," *Legal Times*, July 4, 1988, page 1.

Page 216. *. . . Brandeis University held a debate . . .* Interviews with scientists at the debate; *The Boston Globe*, April 24, 1991, page 13. Also see Bernard Davis' letter to the editor, *The Boston Globe*, May 8, 1991, page 22.

Page 216. *Meanwhile, the impact . . .* Interview with Zinder.

Page 217. *He also hired . . . The Boston Globe*, April 24, 1991, page 13; *Scandal: The Culture of Mistrust in American Politics*, by Suzanne Garment, Times Books, 1991, page 165; "Baltimore Throws in the Towel," by David P. Hamilton, *Science*, May 10, 1991, page 768.

Page 217. *In early May . . .* "Dr. Baltimore Says 'Sorry,' " *Nature*, May 9, 1991, page 94.

Page 219. . . . *"very little admission"*. . . "Baltimore Apologizes for Defense of Research Paper With Bad Data," by Philip J. Hilts, *The New York Times*, May 4, 1991, page 1.

Page 219. *Gilbert later told* . . . "Baltimore Throws in the Towel," by David P. Hamilton, *Science*, May 1991, page 768.

Page 220. *"To make an error* . . ." "The End of the Baltimore Saga," *Nature*, May 9, 1991, page 85.

Page 220. *The article was O'Toole's response* . . . "Margot O'Toole's Record of Events," *Nature*, May 16, 1991, page 180.

Page 222. *Needless to say* . . . "Baltimore Declares O'Toole Mistaken," *Nature*, May 30, 1991, page 341.

Page 226. *Hadley later explained* . . . Hadley interview.

Page 227. . . . *Wortis, Huber, and Woodland responded* . . . "Opinions from an Inquiry Panel," *Nature*, June 13, 1991, page 514.

Page 227. . . . *O'Toole could not let the issue go.* "O'Toole Re-challenges," *Nature*, June 27, 1991, page 692.

Page 228. . . . *vindication for Stewart* . . . "Analysis of a Whistle-blowing," by Stewart and Feder, *Nature*, June 27, 1991, page 687.

Page 230. *Sharp later explained* . . . Sharp interview.

Page 231. . . . *historically an institution* . . . "The Odd Couple," by David Warsh, *The Boston Globe Magazine*, December 1, 1991, page 16.

Page 231. . . . *the NIH whistle-blower* . . . Stewart interview.

Page 232. *Eisen told the Harvard scientists* . . . Transcript of the June 4, 1991, meeting of Eisen and the Harvard scientists.

Page 236. *"Details that could throw doubt* . . ." *"Surely You're Joking, Mr. Feynman!" Adventures of a Curious Character*, by Richard Feynman, as told to Ralph Leighton, Bantam Books, 1989.

Chapter Seventeen

Page 240. . . . *civilized exchange of letters* . . . "More Views on Imanishi-Kari," *Nature*, July 11, 1991, page 101.

Page 240. . . . *Ptashne first described how two other related papers* . . . "Quantitative Analysis of Idiotypic Mimicry and Allelic Exclusion in Mice with a Mu Ig Transgene," by Satyajit Rath, Jeannine Durdik, Rachel M. Gerstein, Erik Selsing, and Alfred Nisonoff, *Journal of Immunology*, September 15, 1989, page 2074; and "Isotype Switching by a Microinjected Mu Immunoglobulin Heavy Chain Gene in Transgenic Mice," by Jeannine Durdik, Rachel M. Gerstein, Satyajit Rath, Paul F. Robbins, Alfred Nisonoff, and Erik Selsing, *Proceedings of the National Academy of Sciences USA*, April 1989, page 2346.

Page 241. *". . . scientific Watergate." Nature*, July 11, 1991, page 101.

Page 241. . . . *distinctly different perspective. Nature*, July 11, 1991, page 102.

Page 242. *In the July 18 issue* . . . "Responsibility and Weaver et al.," *Nature*, July 18, 1991, page 183.

Page 245. . . . *"Dear Paul" letter* . . . "Open letter to Paul Doty," *Nature*, September 5, 1991, page 9.

Page 246. *"He even retracted his retraction* . . ." "David Baltimore's Final Days," *Science*, December 13, 1991, page 1578.

Page 246. *". . . great name recognition."* Zinder interview.

Page 246. . . . *"unique opportunity"* . . . "Noted Scientist and Staff Leave Rockefeller U.," by William K. Stevens, *The New York Times*, August 1, 1991, page B4.

Page 247. *Then, on October 8* . . . "Second Research Group in a Year Is Leaving Rockefeller University," by John Noble Wilford, *The New York Times*, October 9, 1991, page B8.

Page 247. *". . . readers have seen the essence . . ."* "Baltimore's Defense," *Nature*, October 10, 1991, page 484.

Page 248. . . . *they met October 17* . . . *Science*, December 13, 1991, page 1578.

Page 251. Nature *saw the resignation* . . . "Baltimore Defeat a Defeat for Research," *Nature*, December 12, 1991, page 419.

Chapter Eighteen

Page 252. . . . *a growing awkwardness* . . . Hadley interview.

Page 252. *Healy, who was confirmed* . . . "Is American Science Ready for Bernadine Healy?," by Peter G. Gosselin, *The Boston Globe*, July 7, 1991, page 25; "A New Assertiveness at the U.S. Health Institutes," by Philip J. Hilts, *The New York Times*, September 9, 1991, page 13; "Campus Politician," by Graeme Browning, *National Journal*, October 26, 1991, page 2603.

Page 253. . . . *Cleveland Clinic Foundation* . . . Proceedings of the Oversight and Investigations Subcommittee, August 1, 1991. Also see "NIH Probe: Scientist Made False Statements," by John Crewdson, *Chicago Tribune*, July 28, 1991, page 4; "New Questions in Scientific Misconduct Case Prompt Withdrawal of Health Official," by Philip J. Hilts, *The New York Times*, August 1, 1991, page B6; and "Derailing or Due Process?," by Benjamin Weiser, *The Washington Post*, August 14, 1991, page 19.

Page 253. *"We were congratulating . . ."* Hadley interview.

Page 254. *"I've got to have . . ."* Hadley interview.

Page 254. . . . *Storb refused to resign* . . . Hadley and Healy congressional testimony, August 1, 1991: "Conflict of Interest Is Charged in Inquiry on Research Paper," by Philip J. Hilts, *The New York Times*, June 14, 1991, page D18.

Page 255. *Healy was upset* . . . Healy congressional testimony August 1, 1991.

Page 255. *Robert Lanman, who was also at the meeting* . . . Lanman congressional testimony August 1, 1991.

Page 255. . . . *like a novel*. Hadley interview.

Page 255. . . . *badly written* . . . Healy congressional testimony, August 1, 1991.

Page 256. *Healy did not know* . . . Healy congressional testimony.

Page 256. . . . *"rein you in."* Hadley interview. Also see "OSI Investigator 'Reined In,' " by David P. Hamilton, *Science*, July 26, 1991, page 372.

Page 257. *"I was crying . . ."* Hadley interview.

Page 259. *The hearing ended* . . . Interview with Bernadine Healy.

Page 260. . . . *feeling "profoundly ill"*. . . Hadley interview. Also see "FBI Probes NIH Leaks of Science Fraud Files," by Malcolm Gladwell, *The Washington Post*, March 13, 1992, page 3.

Page 263. . . . *tight-lipped tradition* . . . Larry Stewart and John Hargett interviews.

Page 265. . . . *Geoffrey Garinther* . . . "Researcher Accused of Fraud in Her Data Will Not Be Indicted," by Philip J. Hilts, *The New York Times*, July 14, 1992, page C3.

Page 265. . . . *the "scandal here . . ."* *The New York Times*, July 14, 1992, page C3.

Page 265. . . . *retract his retraction.* *The New York Times*, July 14, 1992, page C3.

Page 265. *". . . not to prosecute . . ."* *The New York Times*, July 14, 1992, page C3.

Page 265. *"I've had no change of heart"*. . . Edsall interview.

Page 265. *"Scientific matters . . ."* Interview with Serge Lang.

Page 265. ". . . *less to do with fraud* . . ." Gilbert interview.

Chapter Nineteen

Page 267. *Dingell defended himself* . . . Dingell interview.

Page 267. *O'Toole is oddly sympathetic* . . . O'Toole interview.

Page 268. *Even James Wyngaarden* . . . Wyngaarden interview.

Page 268. *Bernardine Healy, who . . . Healy interview.*

Page 268. *". . . scientific community's refusal . . ."* "Fraud in Scientific Research: The Prosecutor's Approach," adapted from a Breckinridge Willcox paper presented at the Second International Conference on Research Policies and Quality Assurance, May 6–7, 1991, in Rome, *Accountability in Research*, 1992, page 139.

Page 270. . . . *"us versus them"* . . . O'Toole speech at the Cavallo Foundation reception June 25, 1992.

Page 270. . . . *"kindergarten ethics"*. . . Stewart interview.

Page 271. . . . *little reason to stay.* Baltimore interview.

Page 272. *In the flurry of meetings* . . . See *Responsible Science*, by the Panel on Scientific Responsibility and the Conduct of Research (of the Committee on Science, Engineering, and Public Policy), National Academy Press, 1992, Volume 1.

Page 273. *Many of the problems* . . . See *Impure Science: Fraud, Compromise and Political Influence in Scientific Research* by Robert Bell, John Wiley & Sons Inc., 1992; "Peer Review: Treacherous Servant, Disastrous Master," by Charles W. McCutchen, *Technology Review*, October 1991, page 28; "Biomedical Information, Peer Review, and Conflict of Interest as They Influence Public Health," by Erdem I. Cantekin, Timothy W. McGuire, and Robert L. Potter, *Journal of the American Medical Association*, March 9, 1990, page 1427.

Page 273. *"One of the problems . . ."* *Science & Government Report*, March 15, 1990, page 4.

Page 273. *"If I could make a rule . . ."* Temin interview.

Index

■